DISCOURSE, POWER, AND RESISTANCE

TRANSGRESSIONS: CULTURAL STUDIES AND EDUCATION

Volume 88

This book series is dedicated to the radical love and actions of Paulo Freire, Jesus "Pato" Gomez, and Joe L. Kincheloe.

Cultural studies provides an analytical toolbox for both making sense of educational practice and extending the insights of educational professionals into their labors. In this context *Transgressions: Cultural Studies and Education* provides a collection of books in the domain that specify this assertion. Crafted for an audience of teachers, teacher educators, scholars and students of cultural studies and others interested in cultural studies and pedagogy, the series documents both the possibilities of and the controversies surrounding the intersection of cultural studies and education. The editors and the authors of this series do not assume that the interaction of cultural studies and education devalues other types of knowledge and analytical forms. Rather the intersection of these knowledge disciplines offers a rejuvenating, optimistic, and positive perspective on education and educational institutions. Some might describe its contribution as democratic, emancipatory, and transformative. The editors and authors maintain that cultural studies helps free educators from sterile, monolithic analyses that have for too long undermined efforts to think of educational practices by providing other words, new languages, and fresh metaphors. Operating in an interdisciplinary cosmos, *Transgressions: Cultural Studies and Education* is dedicated to exploring the ways cultural studies enhances the study and practice of education. With this in mind the series focuses in a non-exclusive way on popular culture as well as other dimensions of cultural studies including social theory, social justice and positionality, cultural dimensions of technological innovation, new media and media literacy, new forms of oppression emerging in an electronic hyperreality, and postcolonial global concerns. With these concerns in mind cultural studies scholars often argue that the realm of popular culture is the most powerful educational force in contemporary culture. Indeed, in the twenty-first century this pedagogical dynamic is sweeping through the entire world. Educators, they believe, must understand these emerging realities in order to gain an important voice in the pedagogical conversation.

Without an understanding of cultural pedagogy's (education that takes place outside of formal schooling) role in the shaping of individual identity – youth identity in particular –

the role educators play in the lives of their students will continue to fade. Why do so many of our students feel that life is incomprehensible and devoid of meaning? What does it mean, teachers wonder, when young people are unable to describe their moods, their affective affiliation to the society around them. Meanings provided young people by mainstream institutions often do little to help them deal with their affective complexity, their difficulty negotiating the rift between meaning and affect. School knowledge and educational expectations seem as anachronistic as a ditto machine, not that learning ways of rational thought and making sense of the world are unimportant.

But school knowledge and educational expectations often have little to offer students about making sense of the way they feel, the way their affective lives are shaped. In no way do we argue that analysis of the production of youth in an electronic mediated world demands some "touchy-feely" educational superficiality. What is needed in this context is a rigorous analysis of the interrelationship between pedagogy, popular culture, meaning making, and youth subjectivity. In an era marked by youth depression, violence, and suicide such insights become extremely important, even life saving. Pessimism about the future is the common sense of many contemporary youth with its concomitant feeling that no one can make a difference.

If affective production can be shaped to reflect these perspectives, then it can be reshaped to lay the groundwork for optimism, passionate commitment, and transformative educational and political activity. In these ways cultural studies adds a dimension to the work of education unfilled by any other sub-discipline. This is what Transgressions: Cultural Studies and Education seeks to produce – literature on these issues that makes a difference. It seeks to publish studies that help those who work with young people, those individuals involved in the disciplines that study children and youth, and young people themselves improve their lives in these bizarre times.

Discourse, Power, and Resistance Down Under

Edited by

Mark Vicars
Victoria University, Melbourne, Australia

Tarquam McKenna
Victoria University, Melbourne, Australia

and

Julie White
La Trobe University, Melbourne, Australia

SENSE PUBLISHERS
ROTTERDAM / BOSTON / TAIPEI

A C.I.P. record for this book is available from the Library of Congress.

ISBN 978-94-6209-035-4 (paperback)
ISBN 978-94-6209-036-1 (hardback)
ISBN 978-94-6209-037-8 (e-book)

Published by: Sense Publishers,
P.O. Box 21858, 3001 AW Rotterdam, The Netherlands
https://www.sensepublishers.com

Printed on acid-free paper

TABLE OF CONTENTS

PETER BANSEL

RESISTING AND RE/COUNTING THE POWER OF NUMBER

The one in the many and the many in the one

In this chapter I turn the themes 'power,' 'discourse' and 'resistance' to contemplation of the place of qualitative accounts of experience within contemporary managerialist audit cultures and their preoccupation with quantity, number and counting. I specifically critique and resist the power and status of number and quantity as key indicators of what matters and what counts. I do this through play with and on vocabularies of 'number' and 'counting,' and through laying out an account of my thinking about discourse, narrative, subjectivity and experience. In contesting the privileged status of numerical counting, I elaborate a theoretical, philosophical and epistemological approach to subjectivity and narrated biographical accounts of experience. In so doing, I write-over one form of accounting (numerical) with another (narrative). I also resist recourse to validation by a vocabulary of number – where quantity is a proxy for significance, validity and quality – and emphasise the extent to which qualitative accounts of any 'one' are understood as simultaneously those of 'many.'

My resistance to managerialist and reductionist uses of number for the purposes of audit and validation is extended to certain iterations of discourse and discourse analysis, and to aspects of neoliberalism as a mode of government. I resist over determined accounts of the operation of discourse in the constitution and regulation of the subject, as well as the reduction of the subject to positions within fixed discursive repertoires. In relation to neoliberalism, I resist a preoccupation with number and quantification that places greater value on quantitative data for determining the value of anything and everything; and the neoliberal preoccupation with the individual, individuality and individualism. Where once Margaret Thatcher (an early enthusiast of neoliberalism) announced, 'there is no such thing as society. There are individual men and women' (quoted *Women's Own* magazine, October 31, 1987), I'm suggesting instead that there is no individual. Rather, I give an account of the subject as a co-constitutive relation with others. This is in part, a contemplation on the ethics of numbers and their relation to the accounts we might give of the subjects they describe, measure and quantify.

RESISTING NUMBERS

Numbers, says Rose (1999a, p. 197), have achieved 'an unmistakable power within technologies of government.' Within neoliberal political reason, numbers operate

M. Vicars et al. (eds.), Discourse, Power, and Resistance Down Under, 1–8.

as diagnostic instruments promising to align the exercise of public authority with the values, beliefs and well-being of citizens. Numbers help constitute and regulate the domains upon which government operates, mapping the internal characteristics of the spaces of population, economy and society. Numbers are political – because political judgements are implicit in the choice of what to measure, how to measure it, how often it should be measured and how to interpret and represent the results. Espeland and Stevens (2007) make the point that measurement doesn't simply account for the social worlds it depicts, but actively intervenes in them – leading people to think and act differently. Acts of measurement both create and reproduce social boundaries and categories, and ameliorate complexity through the production of distinctions between categories of people. Measurement turns all difference into quantity – and through this process qualitative differences become quantitative ones. The turning of qualities into quantities creates new objects of measurement and new relations among them.

And yet, as Rose (1999a, p. 198) points out, whilst numbers seem indispensable to politics, they also appear to depoliticise whole areas of political judgement. That is, they redraw the boundaries between politics and objectivity by purporting to act as transparent technical means for making judgements, prioritising problems and allocating resources. Numbers, and the professional and specialist knowledges and professional techniques that are associated with them, become implicated in the creation of specific domains – discursive, technical, practical, personal, political and social. In this way numbers, and the technical expertise associated with them, can come to dominate political debate and public imagination. This gives rise to the quantification – that is 'the production and communication of numbers' – of all aspects of social phenomena, as well as the proliferation of new regimes of measurement and technologies of counting (Espeland & Stevens, 2008, p. 401). In these ways, all social phenomena have a primarily metrical relationship to one another. But how else might the relationship between social phenomena be accounted for? What counts as evidence? How exhaustive (and exhausting) does our counting need to be to be credible?

RE/COUNTING THE SUBJECT

Questions of transparency, validity, objectivity and truth form the boundaries of what passes as evidence. There is a constitutive relationship between quantification and practices of government that is intensified within democratic neoliberal government in capitalist economies, where everything that counts is reduced to a number. These numbers are given political and social currency that is invariably converted into a monetary value that can be measured and traded in a globally competitive market economy. This leads to the reification of accounting as a technical practice that produces legitimacy, shapes preferences, organisational structures and practices, as well as the forms of visibility which support and give meaning to decision making (Power, 2003). Scientistic assumptions behind measurement and statistical sampling privilege algorithmic technologies that idealise calculation and enumeration as a series of logical steps through which:

facts can be laid bare; problems can be identified and solved; and decisions about populations and individuals can be made.

As the premises behind the numbers disappear, the decisions, facts, truths and realities they give rise to become ever more normalised, naturalised, obvious and self-evident. Indeed, numbers stand in the place of what is measured, and what is measured becomes a spectral presence in the tabulated metrics they are reduced to. The real becomes co-extensive with what is measurable. Measurement narrows our appraisal of value and relevance to that which can be easily measured, and does this at the expense of other ways of knowing. To count one must be counted. What I am interested in here, as an act of resistance to this numerical reductionism, is a 'different way of thinking about accountability, one that focuses on accounts of experience' and draws on 'qualitative, multiple and local measures' (Shore & Wright, 1999) to vivify rather than validate our accounts of the social.

Numbers have a performative force; they are persuasive and authoritative and they accumulate and link users who have investments in the numbers. Espeland and Stevens, drawing on Actor Network Theory, argue that the authority of numbers, like that of scientific facts more generally, 'depends on establishing networks among objects and humans that become so sturdy that they are no longer disputed or subject to disassembly' (Espeland & Stevens, 2008, p. 420). What, I ask, if we were to foreground the constitutive networks of relations between and among subjects, and give accounts of them as multiplicities that do not need to be (indeed cannot be) quantified or metricised?

In resisting the over-determination of the place of number in the constitution and regulation of the subject and the social, I'm interested in a different form of accounting – the accounts that we might give of ourselves and of each other. As Kamuf (2007) suggests, narrative accounting and computational counting are commonly thought to stand in rough opposition to each other. The former is pejoratively dismissed as subjective accounts of experience and the latter privileged as what counts as objective, fact, evidence and proof. Numbers apparently (and unlike people) 'do not lie'; they tell the real story which, Kamuf reminds us, 'is the story of no story to tell' (2007, p. 252), no account to be given, only an accounting to be undertaken. Yet, if practices of calculation reduce everything to a number, the number I am interested in is the number one; where that one is always more than singular.

THE SUBJECT AS MORE THAN SINGULAR

I am specifically concerned with giving a post-positivist account of the subject as imbricated in co-constitutive relationships with others. Such an account informs my theoretical orientation to the research practice of conducting and analysing biographical narrative interviews. My concern in this chapter is not so much with methodology – how to conduct and analyse narrative interviews – but with epistemology – what it might be possible to know about the subject, practices of subjectification and the social world from engagement with a single narrative account of experience (for an example of this, see Bansel & Davies, 2010).

Post-positivist accounts of subjectivity and experience typically foreground the constitutive force of language or discourse in the making of the subject. Subjectivity, understood as positioning within discourse, is constituted and regulated through the discursive practices that characterise historically specific practices of subjectification. Yet, like Henriques et al. (1998), I recognise the dangers of discourse determinism that imply that people are mechanically and unilaterally positioned in similar and stable discourses with similar and stable effects, and little more than exemplars of repertoires deduced from the quantification of patterns in discourse. The 'one' is, in this view, little more than an example of the many.

Whilst I recognise that discursive and normative practices of subjectification have material effects (Foucault, 1978, 1978-9, 1991), I also recognise that subjects are more than an 'effect' of these discourses. Rather, they are subjects 'whose conditions of emergence can never be fully accounted for' (Butler, 2005). That is, they can never be fully described, reduced to, or amassed as indictors, taxonomies or positions that are at once numerically cumulative and reductionist. I emphasise the multiple, complex and irreducible intersections between multiple discourses and discursive practices, formations and positions through which subjects are constituted. These multiple intersecting discourses are not mere accumulations of single discourses that might be examined in isolation, but rather, necessarily understood in terms of the constitutive networks and relations through which they are mobilised into repertoires that adhere to each other in the formation of a constellation of meanings, possibilities for action, or matrices of intelligibility (Butler, 1990). I am, then, giving an account of an irreducible subject; that is, a subject constituted through discourse and discursive practices, but not totally accounted for, reducible to, or fully described by them. This is a subject who resists codification or counting as a singularity.

I also resist the centrality of the individual in neoliberalism as a mode of government. As articulated by Nikolas Rose (1996, 1999, 1999a), neoliberalism is a mode of government through which subjects are constituted as autonomous and rational agents of democracy; subjects able to determine the course of their lives and identities through the choices they make. This is described by Rose as both a fiction (that is a fiction of an autonomous agentic self) and as a biographical project of self-realisation. I work with this idea of a fictive biographical project of the self, and emphasise that this project, and the self it produces, is *relational* rather than rational and autonomous. I am also interested in the extent to which it might be understood as fictive. I think about this in two ways. First, I argue that the rationality and autonomy of the biographical self is itself a fiction. Second, I take up the trope of 'fiction' and recuperate it from its other, 'fact,' in order to resist the reduction of the subject to a series of numbers or percentages, a compilation of facts or factors, an inventory of characteristics or even a repertoire of discourses or discursive practices.

I emphasise that the 'I' who speaks is not a 'one,' but rather a 'Fictive-I' who is constituted relationally through engagement with multiple others. I want to dislodge the singularity of this 'Fictive-I' and reiterate it as the many. This is strategic for at least three reasons: one, a preoccupation with a political and ethical relation of the self to the other; two, as resistance to the primacy of an autonomous individual within neoliberal practices of government; and three, to disrupt iterations of discourse analysis that articulate the one as simply an exemplar of the many, rather than understanding the one to also *be* the many, such that the many might be read in and from the 'one.'

This 'one' is relational and multiple: there is no 'I' without a 'you,' and neither this 'I' nor 'you' are singular (Ricoeur, 1994; Cavarero, 2000; Butler, 2005). My preoccupation with relationality is directed towards thinking about a biographical narrative account of experience as more than simply an account of oneself, but rather, as an account of the times, places and practices through which the 'I' is constituted. Rather than think of the one as an example of the many, I seek to articulate the simultaneity of the one in the many and the many in the one. In this play with numbers I resist the fantasy or conceit that measurement provides privileged or exclusive access to the universal or to the real. I resist too, any call to the real as a stable, durable and persistent ontology that can be known through acts of quantification.

THE ONE IN THE MANY AND THE MANY IN THE ONE

I have emphasised that the apparently singular embodied subject, the one who gives an account of her or himself, is a co-constitutive and co-extensive relation with others. This implies a particular iteration of the relation between the one and the many, contests the ways in which the one is understood as an example of the many, and foregrounds instead, the ways in which the many might be understood from an engagement with the one.

In foregrounding a relational and multiple subject, and a relation between the one and the many, I draw on Nissen's (2005) account of collective subjectivity. Nissen's account of subjectivity foregrounds 'the situated activities or practices that are emergent from particular instances of dynamic reciprocal constitutive relations among participants and collectives' (2005: 154). Within these co-constitutive relations both specific embodied subjects and the social body are understood 'as *particular subjects*' (Nissen, 2005, p. 154, original emphasis). For Nissen this is crucial, 'since it establishes the collective as a singular object for … study' (2005, p. 154). In this way, Nissen articulates the collective as a particular subject, and such a subject is understood in terms of relations of participation. Particular collectives are understood as singular subjects who participate in larger practices, and themselves perform reflexivity 'in the explicit shape of a we,' but also in the ongoing regulation of individual activity (Nissen, 2005, p. 156). I am interested in this movement between the one and the many, where the 'one' is understood as collective and the collective understood as 'one.' The subject of research might then be understood as simultaneously collective and singular; the many a reading of the one, and conversely, the one a reading of the many. As Judith Butler suggests, the 'one' who speaks is never a 'one,' but a performance of a 'one' who is, in part, 'the enigmatic traces of others' (Butler, 2004, p. 46).

In thinking about these enigmatic traces of others I draw on Deleuze and Guattari's (2004) concept of collective assemblages of enunciation. The idea of a collective assemblage accounts for the social character of utterances. It also incorporates the pluralised voices of others, the many voices within a single voice, and renders a notion of individual enunciation redundant. Further, if, as de Certeau suggests, 'memory is a sense of the other' (1988, p. 87), then the accounts that we give of ourselves are more than accounts of ourselves: they are also accounts of the collective relations and memories through which any 'I' might give an account of her- or him-self. 'Why,' asks Ricoeur 'should memory be attributed only to me, to you, to her or to him, in the singular ... And why could attribution not be made directly to us, to you in the plural, to them?' (2006, p. 94).

Though particular memories might emerge at the site of a particular subject, 'their associations extend far beyond the personal. Rather, they spread into an extended network of meanings that bring together the personal with the familial, the cultural, the economic, the social and the historical' (Kuhn, 1995, p. 233). It is possible then to read the narrated account of a particular embodied subject as a relation among other subjects, discursive formations, historical events, family histories, structures of feeling, and relations of class, gender, ethnicity, sexuality and so on.

The memory or narrative of one subject is, then, a trace of the memories or narratives of other subjects. There is a movement I am articulating here, between the one and the many, the singular and the collective. In place of reading the one as an exemplar of the many, I am more interested in the possibility that the many might be read from the one: a 'one' who is an assemblage of the many. The narrated 'I,' is a heteronomous subject who feigns autonomy. As Secomb reminds us, speech secures its efficacy 'not through the single utterance of a lone speaker but through repeated citations by a chorus of speakers' (Secomb, 2007, p. 152).

The accounts we give of ourselves are, in this sense, understood as polyvocal, or rhizovocal (Youngblood Jackson, 2003). They are the eruption and articulation of many voices in 'one.' Attention to these many voices emphasises the constitutive technologies of government and self through which the appearance, narratability and intelligibility of subjects is accomplished. It is, I propose, through attention to this relationality that we might understand any particular subject as simultaneously an iteration of the many, and the many as an iteration of any particular one.

Cavarero (2000) emphasises that we are, of necessity, exposed and vulnerable to one another. We are not, she says, separate, isolated individuals. Rather, we exist both for the other, and by virtue of the other: without 'you,' says Cavarero, my own story becomes impossible. I am interested not simply in the idea that the accounts that we give of ourselves depend on a constitutive relation between one *and* another, but referencing Paul Ricoeur, emphasise that these accounts depend on a relation of oneself *as* another (Ricoeur, 1994). The concept of oneself *as* another 'suggests from the outset that the selfhood of oneself implies otherness to such an intimate degree that one cannot be thought of without the other' (Ricoeur, 1994, p. 3). To speak, then, of oneself *as* another, is to recognise that we are imbricated in networks of relations and that autonomy is an impossible fiction. As

Deleuze entreats, 'Be yourselves ... it being understood that this self must be that of others' (2004, p. 158).

We are from the outset, given over to the other, and given over to the other, we are 'beyond ourselves, implicated in lives that are not our own' (Butler, 2004, p. 22). This relationality, or 'field of ethical enmeshment with others,' produces a 'sense of disorientation for the first-person,' in which we are always 'more than, and other than, ourselves' (Butler, 2004, p. 25). The subject, as more than itself, is 'a piling up of heterogeneous places' (de Certeau, 1988, p. 201); a subject 'whose epochs *all* survive in the *same* place, intact and mutually interacting' (de Certeau, 1988, p. 202, original emphasis). These mutually interacting subjects, times and places are brought together in any narrated account that we give of ourselves. That they '*all* survive in the *same* place,' intimates the possibility of a 'one' who is simultaneously everyone, everywhere and everything; a one who is many, and a many who are one.

ADDING THINGS UP

In this post positivist, philosophical reflection on narrated accounts of experience I have hoped to re-appropriate the concept of accounting from its association with computation, numbers, metrics and transparency by: making more transparent the ways in which the politics of number operate within neoliberal mentalities of government; and substituting 'narrative accountancy' for 'numerical accountancy.' I have suggested that the singular and autonomous subjects of modernist and neoliberal discourse is an impossible fiction, and that we always were, and will continue to be, multiple and relational.

I have also foregrounded possible escape from the postulate of singularity through an account of the subject as composed of co-constitutive relations among others. I have suggested that any single account of experience is always an account of other times, places, subjects and practices, and is in this sense a collection of multiple heterogeneous elements. Further, these heterogeneous elements can be re-constituted from within any single account, and reassembled as accounts of the social or the collective. Against a rationality of quantity, I suggest that the qualitative accounts we give of ourselves and each other do not require the imprimatur of any number other than the number one; which was always already a multitude. This, is finally, a question of the ethics of number, of who or what counts in a social order that fetishises individuality and quantity.

REFERENCES

Bansel, P., & Davies, B. (2010). Through a love of what Neoliberalism puts at risk. In J. Blackmore, M. Brennan, & L. Zippin (Eds.), *Re-positioning university governance and academic work*. Rotterdam: Sense Publishers.

Butler, J. (1990). *Gender trouble: Feminism and the subversion of identity*. London/New York: Routledge.

Butler, J. (2004). *Precarious life: The powers of mourning and violence*. London: Verso.

Butler, J. (2005). *Giving an account of oneself.* New York: Fordham University Press.

Cavarero, A. (2000). *Relating narratives: Storytelling and selfhood.* London: Routledge.
de Certeau, M. (1988). *The practice of everyday life.* Berkeley/Los Angeles/London: University of California Press.
Deleuze, G. (2004). *Difference and repetition.* London/New York: Continuum.
Deleuze, G., & Guattari, F. (2004). *A thousand plateaus; Capitalism and schizophrenia.* London: Continuum.
Espeland, W., & Stevens, M. (2008). A sociology of quantification. *European Journal of Sociology*, 49, 401-436.
Foucault, M. (1978). On governmentality. *Ideology & Consciousness, 6*, 5-22.
Foucault, M. (1978-1979). *The birth of biopolitics: Lectures at the College de France 1978-1979.* In M. Senellart (2008) (Ed.), *Michel Foucault, The birth of biopolitics: Lectures at the College de France 1978-1979.* Hampshire/New York: Palgrave Macmillan.
Foucault, M. (1991). *Discipline and punish. The birth of the prison* (trans A. Sheridan). London: Penguin.
Henriques, J., Hollway, W., Urwin, C., Venn, C., & Walkerdine, V. (1998). *Changing the subject: Psychology, social regulation and subjectivity.* London: Routledge.
Kamuf, P. (2007). Accounterability. *Textual Practice, 21*(2), 251-266.
Kuhn, A. (1995). *Family secrets: Acts of memory and imagination.* London/New York: Verso.
Nissen, M. (2005). The subjectivity of participation: Sketch of theory. *International Journal of Critical Psychology: Neoliberal Subjects, 15.*
Power, M. (2003). *The audit society: Rituals of verification.* Oxford: Oxford University Press.
Ricoeur, P. (1994). *Oneself as another.* Translated: K. Blamey. Chicago/London: University of Chicago Press.
Rose, N. (1996) Governing "advanced" liberal democracies. In A. Barry, T. Osborne, & N. Rose (Eds.), *Foucault and political reason.* Abingdon: Routledge.
Rose, N. (1999). *Governing the soul* (2nd Edition). London: Free Association Books.
Rose, N. (1999a). *Powers of freedom: Reframing political thought.* Cambridge: Cambridge University Press.
Secomb, L. (2008). Words that matter: Reading the performativity of humanity through Butler and Blanchot. In B. Davies (Ed.), *Judith Butler in conversation.* New York/London: Routledge.
Shore, C., & Wright, S. (1999). Audit culture and anthropology: Neo-liberalism in British higher education. *The Journal of the Royal Institute, 5*(4), 557-574.
Youngblood Jackson, A. (2003). Rhizovoality. *QSE: International Journal of Qualitative Studies in Education, 16*(5), 693-710.

Peter Bansel
School of Social Sciences and Psychology
University of Western Sydney, Australia

IRIS DUMEDEN

THE RESEARCH ASSISTANT

Invisible and silenced, exploited and disposable

> Author's note: This is a 'true' story – that is, a story that 'fit[s] in the space between fiction and social science, joining ethnographic and literary writing, and autobiographical and sociological understanding' (Ellis, 1993, p. 711). This story is also what Ellis calls 'an introspective case study in emotional sociology' in which I position 'myself reflexively as both narrator (author) and main character of the story' (p. 724). This form of narrative allows researchers to acknowledge and give form to their own emotional experiences, and to reclaim and write their own lives (p. 711). The goal is to encourage readers 'to "experience an experience" that can reveal not only how it was for me, but how it could be or once was for them' (p. 711).

Doing doctoral research is a pretty lonely enterprise. That I had chosen to do an ethnographic study of just one student – tutoring and mentoring him intensively during his final year of schooling – made my research study even more limited, in terms of meeting and interacting with other people. Coupled with my role as fulltime carer for my elderly mother, my personal world at the beginning of my PhD candidacy revolved mainly around my research student-participant, my PhD supervisor, and my elderly mother. Of course, I did have friends with whom I went out to dinner occasionally, but my day-to-day social contact was limited. It was thus with great enthusiasm that I accepted a research assistantship role offered to me by one of the lecturers at my university, partly for the close compatibility of her research projects with my doctoral research, but mostly for the expanded social contact that would be available to me through her research projects, and lastly, for a chance to learn how to 'do research' from somebody who was more senior that I was in these matters. This has always been my motivation in accepting all other research assistantship positions.

I came to be doing a doctoral degree, not after long planning and strategising, but more as an opportunity that presented itself, as many such opportunities did in my relatively new career in the field of education. Before this, I had worked in Information Technology for more than twenty years, working variously as systems programmer, applications programmer, systems designer, data base analyst, and project manager in Australia and overseas. In 2003, I came back to Australia to become my mother's fulltime carer, as she had had a stroke. After several months, I felt I needed to get out of the house to do something else other than fulltime carer's work. And so I responded to an advertisement in the local papers, asking

M. Vicars et al. (eds.), Discourse, Power, and Resistance Down Under, 9–19.

for volunteers to tutor newly-arrived refugees in English. It asked for only a one-hour-a-week commitment. I thought that was manageable, and thus began my new career in literacy and teaching. I soon asked for more hours of volunteering and eventually ended up as a volunteer-tutor in the classroom, helping out in pre-literacy classes for refugees. When I decided to get teaching qualifications, I tentatively started with the Cambridge CELTA.[i] After this short course, I knew I wanted to be an ESL[ii] teacher, and so I undertook a Postgraduate Diploma in TESOL,[iii] which expanded into a Master in Adult Education, and before I knew it, I was doing a PhD. Thus, I arrived in the PhD program feeling inadequate in terms of my research skills and eagerly jumped into every opportunity that allowed me to work with more senior researchers and learn from them. In the period of my PhD candidature from 2008 to 2010, I worked as a Research Assistant in five research projects. The events described in the rest of this paper centre on two projects that were run simultaneously by three women lecturers from one university.

From my experience, a Research Assistant position is not fixed or clearly defined. The tasks and expectations depend on the skills and talent that the person brings to the role, as well as the extent of her willingness to throw herself into the role. Research Assistants are what Reay (2000, 2004) describes as 'contract researchers' – that is, the manual labourers in the research team who do the invisible domestic labour of academic research. And, just like domestic labour, contract research is most often done by women. Reay's experience of contract research was in fixed-term appointments, funded by university departments from research grants. However, my experience of contract research was more precarious – I was usually offered a certain number of hours at the start of the project, often much less than a hundred hours. And while this was usually extended, the funded researchers[iv] who employed me tended to add on new hours and requests, so my employment conditions were always tentative, short-term and changing. For the two research projects that I report on here, I was initially hired for 36 hours in the first project, and 97 hours in the second. I was eventually given more hours and more tasks, which totalled to 176.5 hours for the two projects, and for which I was paid a total of $5559 for six months' work. This monetary compensation is appalling, considering the amount of work that I did in these two projects, which I detail below. However, this was not a major concern for me when I took on these research assistantship roles, as I was in a relatively stable financial position (I had paid off my home and had no debts) and had a PhD scholarship. Nevertheless, the financial remuneration needs to be borne in mind in this tale.

The attraction of the two projects for me was the opportunity to work with three more senior researchers, and I thought we all shared the same epistemological framework of social justice, advocacy for the disadvantaged and marginalised, and fidelity to feminist ideals. I was excited to be an insider and to be able to see up close – and participate – in the workings of a research team. I envisaged this a change from my lonely doctoral existence and an exciting opportunity to collaborate with women researchers from whom I hoped to learn how to do 'proper' research. Of particular interest to me was how data would be analysed: I anticipated the discussions about suitable theoretical frameworks that would come after the laborious data collection phase. I also assumed that I would be invited to write collaboratively with these women. (This assumption was made on the basis of my experience as Research Assistant to one of these women: we had co-presented our findings from two projects as paper presentations in two international conferences in the UK.) I was excited about the learning experiences that these two new projects promised, and was determined not to miss a moment of learning. I threw myself into these projects and to say 'body and soul' is not an exaggeration. I was unwavering in my goal to see these two projects through to completion by showing my commitment to the projects and to the senior researchers whom I assumed would appreciate my dedication and would reciprocate in some way.

Even before the grant proposals for both research projects were approved, I was asked to familiarise myself with the projects and with the ethics application/approval process for the university in order to be able to jump into the role quickly, and write the ethics applications, plain language statements, consent forms, as well as the questions for the survey instruments that formed a major part of both projects. This was necessary because the lead researcher was going on holiday, and she wanted the ethics application forms signed and submitted before her holiday started. When finally the grant money was allocated, I had two weeks to complete two university ethics applications and one for the Department of Education and Early Childhood Education – which I did by working day and night, sometimes up to midnight (see the email dated 15 June 2010 below). I hand-delivered the printed and collated forms to the home of the lead researcher on the Thursday evening just before her holiday, so she could sign them and they could be submitted on the Friday.

The series of emails[v] below (presented in chronological order) indicate what I was asked to do, in the role of Research Assistant for these two projects. Apart from the ethics application, I was also asked to '*manage*' both the project and the researchers (their tasks, their schedules, their '*education*'). As enticement to persevere in the '*laborious*' tasks ahead, the lead researcher promised that I would find the project '*engaging*' and the data '*very interesting.*'

Hi Iris

Welcome to your first official day on our projects! Could you begin an ethics application for the first project please?

(7 June 2010)

Iris, can you manage us to review the survey?

(10 June 2010)

Iris, are you now in a position to develop the plain language statements and consent forms for the school? Also, could you make up a timeline for us including when people are here or away and school holidays so we can begin scheduling – gaining consent, data collection, etc. – before the holidays?

(14 June 2010)

More thanks. You are traveling at the speed of light! Hope you are finding the idea of the project engaging. The ethics application process is always laborious, but I think the data is going to be very interesting.

(15 June 2010)

Hi Iris,

I wonder if you could do a best case scenario timeline tracking both research projects including ethics approvals, distribution and collection of consent forms, data collection events, teacher focus groups, analysis block of time, reporting obligations, fortnightly team meetings, etc. This would be a first draft – something to respond to.

(2 July 2010)

Hi Iris,

Perhaps you could send an email around asking for a suitable time for a phone meeting before [researcher] goes away again, and a good day to schedule fortnightly meetings.

(5 July 2010)

Iris, can you educate us regarding monitoring input? Also, email [participant] and ask her which days are good for us to work in the classrooms and meet after school.

(20 July 2010)

Agenda for the team meeting:

12noon – Iris joins us

- Finalise plain language statements and consent forms for ethics resubmit
- Students' survey development
- Teachers' survey development
- Parents' survey development
- Data collection/analysis
- Scheduling

- Other

(28 July 2010)

A major part of these two research studies were the online surveys: a total of seven for both projects. I was asked to lead the design of the surveys, which involved the wording and sequencing of the questions. I was then required to turn these surveys into online form using *SurveyMonkey.*[vi] During the data collection phase, I monitored the survey responses, identified gaps and required information, developed additional questions and removed unproductive items. For the first project, I sent out the invitations and survey links to participants, sent weekly reminders and responded to questions and complaints from participants.

Eventually, I was also asked to undertake analysis work on the survey responses:

> Iris, I wonder if you are in a position to agree to do some preliminary data analysis on the surveys from the two projects?
>
> (1 Sept 2010)

For the second project, in addition to the online surveys, I was also asked to make school visits and complete the tasks set out in the following email:

> Iris, would you be able to do a school visit either the first or second week of term – at a time convenient to you and the teachers? This would involve meeting with the teachers, collecting and making notes on their planning, and taking photos of teacher and student artefacts – much as you did with the previous student project.
>
> (1 Sept 2010)

I made three school visits over a six-month period. During these visits, I talked to the teachers and students, observed classes, watched the students work on projects, took photos and videos of the teachers and students at work and collected other artefacts related to the project. More analysis work was requested – on both the survey results and the collected artefacts:

> Thanks for agreeing to undertake some further analysis. Could you please begin a PowerPoint of the data from the survey when it's convenient? I have more visual data. It would be great if you could put these into PowerPoint presentations as you did for the first school visit.
>
> (15 Sept 2010)

As more and more of these analysis requests came my way, I sent off an email asking about the nature of the analysis I was being asked to do:

> Regarding the data from the project survey – what kind of analysis are we looking at here? Also, why PowerPoint?
>
> (27 Sept 2010)

I promptly received the following reply:

> I was hoping to have all graphs from the first project's survey and raw comment data in a PowerPoint so we can adapt as necessary into a series of PowerPoints we will use. We will do the analysis. If you could prepare this before October 15th, that would be great.
>
> (27 Sept 2010, emphasis added)

In response to the above email, I sent off a *PowerPoint* presentation containing a summary of the results of the survey from the first project – in the form of pie charts, bar graphs and tables showing 'before' and 'after' trends. Within the same day, I received the following series of emails, praising my efforts and asking for summary *PowerPoint* presentations for the second research project:

> Iris is developing sensational 'raw data' Powerpoints for us.
>
> These are just terrific. A brilliant start for us.
>
> I have transferred data from the second project onto a memory stick and will post it to you tomorrow. There are teacher and student artefacts, as well as classroom photos. Perhaps start with a large, or maybe three smaller, comprehensive Powerpoints. See what you think.
>
> (Several emails, all dated 27 Sept 2010, emphasis added)

When I received the above emails, I began to discern a change in the tone of the emails, with a marked use of '*you*' versus '*we*' and '*us*' (see highlighted portions). I began to feel uneasy at this point, as I recognised that there was an assertion of distance between the researchers and me. As well, I seemed to be increasingly positioned as the hired help, rather than as a member of the team. Moreover, noting that my analysis work was being called preliminary ('*a brilliant start for us*') and the output of my analysis described as '*raw data,*' I was eager to attend the senior researchers' analysis sessions ('*We will do the analysis*'). I asked to join their meetings in order to watch them '*do the analysis,*' but was told:

> I am sorry to say that we worked on our survey analysis last Friday. I didn't read your email until after that.
>
> (18 Oct 2010)

I was used to receiving several emails a day, as well as receiving prompt replies to my emails. Therefore, this reply that my email was not read until it was too late to have me join their analysis meeting increased my feelings of unease. I also wondered how it was possible to complete the '*survey analysis*' in just one day. At this point, I was still included in the email exchange between the researchers, so I knew when their meetings were scheduled. I asked several times to join the analysis meetings, but I sensed increasing reluctance to include me:

> You can attend the meeting on November 8th. After that, we will make a decision as to whether we think it would be useful for you to attend the 17th.
>
> (24 Oct 2010)

But a few days before the November 8th meeting, I received the following email:

> We have other matters to discuss first and you are most welcome to join us at noon.
>
> (5 Nov 2010)

I knew the meeting was to start at ten in the morning so that '*to join us at noon*' meant that I would miss most of the analysis discussions. Thus, I interpreted the November 5th email as their polite way of saying, 'Please do not ask again.' After this, further emails from me requesting to join them in future analysis meetings went unanswered. As the email exchange between the researchers and me reduced to a trickle, I began to understand that I was no longer part of the team. After months of constant work, immersed in the thick of data collection and analysis, obliged to find the work I was doing '*engaging*' and the data '*interesting*' and therefore to put my heart into the '*laborious*' tasks, I suddenly became invisible and unwanted, excluded even from email. I was devastated.

I went to the university department's Associate Head of School for Research to complain about my exclusion and exploitation. She called us all to a meeting in her office and so I was optimistic about regaining my place in the research team. But, in stark contrast to the warm and collegial atmosphere of my previous meetings with these three women researchers, in this meeting they were cold and uncompromising. The lead researcher read out to me the contract that I had signed when I became the Research Assistant of the two projects, which she said had not specified any analysis or writing tasks. I was also told that the two projects had run out of grant money, and they therefore were no longer in a position to employ me any further. In response, I offered to work without pay in order to see the projects through to completion. (I thought this a viable solution as I had already done it on two previous research projects with one of the researchers – when the grant money had also run out before the data collection and analysis phases were completed.) However, they rejected this offer, protesting that it was '*dangerous*' and that '*the union would come after us*' if they let me work without pay. With no compromise to be reached, the meeting ended rather abruptly and I was left all alone with the Associate Head of School for Research. Aware of my shock at the sudden abandonment, she was initially sympathetic. Eventually, she pointed out to me that academia is really a very individualistic and hierarchical world, and that the best thing I could do was to finish my doctorate. This advice implied that I, too, would soon be doing the same thing as these women researchers had done to me. But it frightened me to think I would eventually become like them and perpetuate such exploitative behaviour as I moved up the academic ladder. This echoed Reay's (2004) own concerns:

> Some of us really want to collaborate in equal teams rather than serve out our time in rigid hierarchies with only the prospect of moving from a position of subservience to one in which we dominate others (p. 36).

I felt my exclusion and marginalisation very keenly. The exclusion was very personal for me and I felt physically, emotionally and psychological ill for several months afterwards. Most of all, I felt shock at the betrayal of trust. I had devoted myself to these research projects in the belief that I was a vital member of the research team, and I had laboured in good faith that these studies were worthwhile endeavours and that important knowledge would be produced. Though I had understood that these projects were also vehicles for the accumulation of Excellence in Research for Australia (ERA) points by the researchers, I had not understood how fiercely they would protect these points, and discard me in the process. I also struggled to understand how the epistemological framework that these women researchers espoused as the underlying motivation for their interest in the projects – that is, fidelity with feminist and social justice ideals, and advocacy for the disadvantaged and marginalised – could be so removed from their own research practices. It was as Reay (2000, p. 19) had lamented: 'The academic gaze, when it does focus on issues of social justice, looks out on the wider social world for evidence of moral and ethical failure, neglecting self scrutiny.'

Reay (1995, p. 368) also observed, while conducting research in a middle-class school, that the practices of 'racism and classism' that characterise a middle-class habitus 'operated beneath a veneer of civility and good manners.' As the excerpts from the emails provided above indicate, the discourse that 'Othered' (Said, 1978) me and my contribution to the research work were often couched in lovely words and fine manners which, for me, obscured the 'Othering' that was actually occurring. Using seemingly benign pronouns that I had glossed over then, but on closer inspection now revealed that they were careful to always refer to me as 'you,' and that 'we' or 'us' *never* included me. Through such discourse, 'I am converted from subject to object, ends to means, from my own woman to their creature' (Reay, 2004, pp. 32-33). This experience also confirms another of Reay's (2000, p. 15) observations: the hierarchical ways of working in academia, wherein research 'signifies as a high-level activity but only when it is embodied in lecturing staff. When it is embodied in the female contract researcher, it becomes reconstituted as a low-level activity.' Thus, in the academic research context, the female contract researcher is not seen as 'an agent in her own right but as a cipher – merely a repository holding research until it is extracted by the real academics' (Reay, 2004, p. 33; see also Hey, 2001; Okley, 1992).

Moreover, in the privileging of the voice of the more senior researchers and the exclusion of mine, I would argue that whatever knowledge that may have been or will be produced using the data collected in these research projects is, as Mauthner and Doucet (2008, p. 973) have argued, a privileging of 'the textual over embodied knowledge.' For example, in my almost daily monitoring of the responses to the online surveys, where I read each respondent's set of answers as if I were reading an interview transcript, I had built up embodied knowledge of the respondents and their responses, which would be quite different from the textual knowledge that the researchers had from their reading of the set of summary graphs, pie charts, and tables that I had produced for them. As well, the researchers had the benefit of the narratives that I presented to them in *PowerPoint* format, but not the embodied knowledge that went into producing those *PowerPoint* presentations they so favoured. As Research Assistant, I had embodied knowledge gained from being present in the field, talking to participants and taking photographs and videos of them while they worked; embodied knowledge gained by studying hundreds of photographs and other artefacts from which I then selected which were to be included in the *PowerPoint* presentations; embodied knowledge gained from the physical act of cutting, pasting, juxtaposing, and captioning these photographs and other visual data; and embodied knowledge from selecting one narrative over other narrative strands that the data may have also presented. I suggest that any knowledge produced without such contextual and embodied knowledge is, as Mauthner and Doucet (2008) assert, merely 'second best.' They note:

> … the distinctive feature of empirical research, in which data are collected in a physical field through embodied research relationships, is that it allows access to non-textual knowledge which can enhance our understanding of the phenomenon in question. We need to ensure that our research practices value and harness this knowledge in its own right alongside textual data and knowledge. (p. 981)

Further, my exclusion from the final phases of research meant that my education as a newcomer to research was partial and incomplete. What I had wanted most from these projects was to join these more senior researchers in action during the inter-subjective and textual construction of knowledge from the collected artefacts. It was this part of the research practice that I had looked forward to, and valued, as the real reward when I accepted the position of Research Assistant on these projects. The $5559 that I was paid was not adequate compensation, but I had rationalised it as a valuable learning experience. And it was – just not in the way that I had envisaged.

Speaking from their experience as research managers in the Australian academic context, Hobson, Jones and Deane (2005) note that there is implicit, if not overt, lack of recognition of the Research Assistant role in the processes of knowledge generation. They argue that, given that academics are time-poor and research is time-intensive, the Research Assistant role is essential to research, but '[p]roblematically, the funded researcher's … capacity to employ a research assistant is an act of self-liberation constituting and constituted in another's subordinate position – the research assistant's' (p. 365). They also note that Research Assistants constituted a 'largely casualized, low-paid and highly gendered workforce that was potentially vulnerable to intellectual exploitation, among other things' (p. 359). In undertaking their study, they asked: 'Put crudely, does the innovation system require an underclass?' (p. 365).

Reay (2000) also notes that contract researchers are employed in the service of their academic colleagues' far more visible academic work. This service relationship, she asserts, raises issues about 'who is entitled to a sense of ownership in relation to research but also who owns the researchers' (p. 15). Based on her own experience as a contract researcher, Reay makes the claim that 'the appropriation of the contract researcher's intellectual labour is normative, routine practice within the academy' (p. 32). She bemoans that the appropriation of one's intellectual labour is a constant hazard for the contract researcher: 'when it does happen you are not even left with a marginalised voice. Your voice becomes subsumed under that of a more powerful academic' (p. 17).

Finally, as Mauthner and Doucet (2008), Hobson, Jones and Deane (2005) and Reay (2000, 2004) have noted, the employment of Research Assistants in a research team is now considered normative. However, as these authors have also pointed out, and as I have tried to show in this account of my experience as a Research Assistant, the role as it is currently constituted is not unproblematic. As Research Assistant, I was invisible and silenced, exploited and disposable. I was hurt by the experience on so many levels – intellectually, emotionally, psychologically and economically – and I would not wish it on anyone.

ACKNOWLEDGMENTS

I would like to thank Julie White for her encouragement and support in writing this chapter. I would also like to thank Keith Simkin and Julie White for their support during the time I was undergoing the personal crisis described in this chapter.

NOTES

[i] CELTA = Certificate in English Language Teaching to Adults.
[ii] ESL = English as a Second Language.
[iii] TESOL = Teaching English to Speakers of Other Languages.
[iv] Funded researchers are most often tenured lecturers who win grants to finance their research projects. The grants pay for research assistants' salaries, among other things.
[v] The emails included here have been fictionalised, a technique which 'allows the author scope to represent absent Others in ways that act as political points of resistance to their enforced absence or silence' (Sparkes, 1997, p. 37). See also Tierney (1993).
[vi] *SurveyMonkey* (www.surveymonkey.com) is an online software tool that allows users to create, design and publish online surveys.

REFERENCES

Ellis, C. (1993). 'There are survivors': Telling a story of sudden death. *The Sociological Quarterly*, *34*(4), 711-730.

Hobson, J., Jones, G., & Deane, E. (2005). The research assistant: Silenced partner in Australia's knowledge production? *Journal of Higher Education Policy and Management*, *27*(3), 357-366.

Hey, V. (2001). The construction of academic time: Sub/contracting academic labour in research. *Journal of Education Policy, 16*(1), 67-84.

Mauthner, N., & Doucet, A. (2008). 'Knowledge once divided can be hard to put together again': An epistemological critique of collaborative and team-based research practices. *Sociology, 42*(5), 971-985.

Okley, J. (1992). Anthropology and autobiography: Participatory experience and embodied knowledge. In J. Okley & H. Callaway (Eds.), *Anthropology and autobiography* (pp. 1-28). London: Routledge.

Reay, D. (1995). 'They employ cleaners to do that': Habitus in the primary classroom. *British Journal of Sociology of Education, 16*(3), 353-371.

Reay, D. (2000). 'Dim dross': Marginalised women both inside and outside the academy. *Women's Studies International Forum, 23*(1), 13-21.

Reay, D. (2004). Cultural capitalists and academic habitus: Classed and gendered labour in UK higher education. *Women's Studies International Forum, 27*, 31-39.

Said, E.W. (1978). *Orientalism: Western conceptions of the Orient.* London: Penguin Books.

Sparkes, A.C. (1997). Ethnographic fiction and representing the absent other. *Sport, Education and Society, 2*(1), 25-40.

Tierney, W.G. (1993). The cedar closet. *International Journal of Qualitative Studies in Education, 6*(4), 303-314.

Iris E. Dumenden
School of Education
Faculty of Arts, Education and Human Development
Victoria University
Melbourne, Australia

LISA GIBBS

CONVEYING THE ROLE OF EVIDENCE IN THE DEVELOPMENT, CONDUCT AND APPLICATIONS OF QUALITATIVE RESEARCH

INTRODUCTION

The push for an evidence-based approach to clinical, policy and service delivery decision making has led to a well developed understanding of the need for studies that meet standards of evidence in order to inform policy and practice change. The literature is replete with papers describing standards of evidence in both quantitative and qualitative research (Daly et al., 2007; Dixon-Woods et al., 2007; Spencer et al., 2003; Public Health Resource Unit, , 2006; Cochrane Collaboration, 2011a; Effective Public Health Practice Group, n.d.; Moher et al., 1995, 1996; Oliver et al., 2005). The growing contribution of Cochrane systematic reviews to the building of evidence is also resulting in increasing sophistication in the processes for determining and developing strength of evidence (Cochrane Collaboration, 2011b) However, there is a gap in the literature in terms of an understanding of the role of evidence in the development, conduct and application of qualitative research.

In this chapter, a four stage representation for the role of evidence in qualitative research is presented which will assist shared understanding in cross-sectoral contexts. It intentionally does not reflect the complexity of qualitative research in terms of theory, epistemology or methodology (Patton, 2002; Hammersley, 1995). Instead it aims to provide a simple tool to assist the early researcher to conceptualise an evidence-informed approach and will also assist with educating the non-academic about the relationship between qualitative research and evidence. This relationship applies to other research methodologies as well but is likely to be enacted differently according to the underlying research paradigms. A case study of research into men's access to chronic illness self management programs is referred to throughout to illustrate the role of evidence at each stage of research.

CASE STUDY

The research being used as a case study in this chapter, "When the whole bloke thing starts to crumble ..." explored the issue of men's access to chronic illness (arthritis) self management programs from a social constructionist perspective (Gibbs & Reidpath, 2005; Gibbs, 2005, 2007, 2008). This consisted of quantitative analyses to confirm sex differences in levels and patterns of service use, and a series of in-depth, semi-structured interviews with 17 men with arthritis and 4

M. Vicars et al. (eds.), Discourse, Power, and Resistance Down Under, 21–26.

partners to gain an increased understanding of the factors affecting men's access to self management programs. The clients and services of Arthritis Victoria were chosen as the setting for this research.

Stage 1: Evidence informs the research question

The initial decision in research about the issue to be studied is generally determined by the need to address a gap in knowledge. Identification of the gap in knowledge is informed by a review of the existing evidence. In a health context this evidence may arise from service data or from the literature, for example identification of a health priority area that is resulting in health inequities or high rates of mortality or morbidity. A broader review of the evidence, such as a literature review, will clarify what is known and what is not known about the research issue. The research question can then be framed to address the gap in the evidence-base.

In the case study example, the issue of men's access to self management programs was first identified through men's lower rates of arthritis self management program utilisation. In recognition of sex differences in the prevalence of the many different types of arthritis, a quantitative analysis of contingency tables and odds ratios was conducted which confirmed that under-representation of men was present even after taking into account sex differences in prevalence (Gibbs & Reidpath, 2005). Under-representation of men as service users was also reported by other chronic illness agencies and in the health literature. However, there was limited evidence about how to increase the accessibility of self management services to men. This gave rise to the research questions:

- Why aren't men participating in existing chronic illness self management programs?
- What kinds of self management programs would men with a chronic illness want to access?

Stage 2: Evidence informs sample selection

Sampling in a research study is determined by what information is required and what methods are being used. Expressed simply, quantitative research seeks limited information from a large amount of people and qualitative research seeks detailed information from a small amount of people. Therefore, the goal of qualitative research is to go straight to the sources most likely to provide detailed, 'rich' information. This is ideally guided by the evidence in terms of identifying who is most likely to be able to provide information about that question and what might the different perspectives be. For example, it may be informed by an unarticulated theoretical perspective such as an expectation that men and women may have different views, or may be defined by existing evidence about the types of people affected by the issue in question. This initial selection of research participants is often referred to as purposive sampling and may be accompanied by progressive sampling as guided by emerging themes in concurrent analysis (theoretical sampling) (Gibbs et al., 2007).

For example, in the case study example, three main sources of variation were sought in the initial purposive sampling, as indicated by patterns in service use and the theoretical literature (Gibbs, 2008):

1) Participation in arthritis self management programs: Given that there was evidence that men were under-represented as service users it seemed reasonable to recruit men who used arthritis self management services and those who didn't (negative case sampling) to learn from the different perspectives of men with different help-seeking behaviours.
2) Social demographics: It was theorised that different socioeconomic and educational backgrounds may influence decision-making about help-seeking behaviours.
3) Self-labelled masculinity: The theoretical literature presented masculinity as a social construction that was likely to influence help seeking behaviours. It postulated that there was a dominant idealised social construction of masculinity but that men enacted their masculinity in different ways in different contexts. Therefore, this study aimed to recruit men who identified with different forms of masculinity to enable exploration of whether this influenced help seeking behaviours.
4) Further theoretical sampling: As part of the concurrent data analysis there was an indication that health behaviours and attitudes may be influenced by the men's age and life stage. The role of female partners was also emerging as a significant influence on decisions and actions to access help. Therefore, further sampling was conducted to ensure a range of ages were represented and to also include men with same sex partners to enable further exploration of those themes.

Stage 3: Evidence informs analysis

Analysis provides an opportunity to examine, code and categorise the data and look for patterns within and across categories (descriptive analysis) (Green et al., 2007). The process of then exploring those relationships and developing a conceptual understanding of emerging themes is enhanced by comparing the findings with data from other sources (triangulation) (Lincoln & Guba, 1985), and with existing research studies, and considering the implications of consistencies and/or conflicts between the findings (conceptual analysis) (Daly et al., 2007). Finally, the relevance of existing theories can be considered to make sense of the findings or to extend the theory to account for any variations in the results (theoretical analysis) (Daly et al., 2007). Each of these different research 'products' reflect the goals of the research and in that sense have equal value but may have different applications (Hammersley, 1995).

In the case study it was found that men's decisions to access chronic illness self management programs were strongly influenced by dominant social constructions of masculinity which constrained help-seeking and health management behaviour

(Gibbs, 2005). However, the restrictive influence of hegemonic masculinity was progressively undermined by the increasing severity of the chronic condition until a crisis point was reached in terms of the severity of the condition or its impact on lifestyle. This resulted in a reformulation or rejection of hegemonic masculinity (Gibbs, 2007). The described conceptual framework was consistent for men from diverse social groupings, although it appeared less prominent in both younger and older men, suggesting that dominant social constructions of masculinity have the greatest influence on health decisions during the middle stage of adulthood when work and family obligations are greatest (Gibbs, 2008). The alignment of the theoretical analysis with existing theories about masculinity (Connell, 1999), and how disabilities impact on masculinity (Gerspick & Miller, 1995), men's concepts of health (Watson, 2000), and research findings about older men's experience of urological conditions (Pinnock et al., 1998), supports this contribution to new knowledge and its application to men and chronic illness beyond the study sampled.

Stage 4. Rigorous findings inform the evidence base

Qualitative research can make a contribution to the evidence in many different ways. Depending on the nature and the quality of the research it can contribute to increased understanding, to the design of quantitative measures, to future research directions, and to inform changes in policy and practice. There are conflicting views about the capacity of qualitative research, given its inherent subjectivity and sampling restrictions, to provide generalisable findings that can provide sufficient evidence to inform changes in policy and practice (Baum, 1995; Schofield, 1990). However, a qualitative study which aims to contribute to theory development, adheres to standards of rigour and demonstrates its alignment with broader theoretical understandings, can also be argued to have applications beyond the sample and context being studied (Daly et al., 2007; Hammersley, 1995).

In addition to a contribution to the international evidence base, the case study findings were used to inform a review of service delivery at Arthritis Victoria and were published in an international health promotion text to provide guidance to health promotion practitioners regarding increasing the accessibility of health programs to men (Gibbs & Oliffe, 2004).

CONCLUSION

Qualitative researchers have an opportunity to educate government, community and funding bodies about the role of evidence in qualitative research. The simple four stage representation outlined in this chapter is intended to provide clarity to this discussion and to demonstrate that evidence has a role in each stage of research development and conduct but may be enacted in different ways according to the underlying theoretical paradigms and epistemologies that govern the research methodologies. Essentially, it confirms that evidence informs each stage of qualitative research and is a key factor in quality and relevance.

REFERENCES

Baum, F. (1995). Researching public health: Behind the qualitative-quantitative methodological debate. *Social Science & Medicine, 40*, 459-468.

Cochrane Collaboration, A Review Group. (2011a). Cochrane effective practice and organisation of care group: EPOC resources for review authors. Available at: http://epoc.cochrane.org/epoc-resources-review-authors, accessed 10 September 2011.

Cochrane Collaboration. (2011b). The Cochrane Library. http://www.thecochranelibrary.com/view/0/index.html, accessed 6 September 2011.

Connell, R.W. (1999). *Men's health: A research agenda and background report.* Canberra: Commonwealth Department of Health and Aged Care.

Daly, J., Willis, K., Small, R., Green, J., Welch, N., Kealy, M., and Huges, E. (2007). Assessing qualitative research in health – A hierarchy of evidence-for-practice. *Journal of Clinical Epidemiology, 60*(1), 43-49.

Dixon-Woods, M., Sutton, A., Shaw, R., Miller, T., Smith, J., Young, B., et al. (2007). Appraising qualitative research for inclusion in systematic reviews: A quantitative and qualitative comparison of three methods. *Journal of Health Services Research & Policy, 12*(1), 42-47.

Effective Public Health Practice Group (n.d.). Quality assessment tool for quantitative studies. http://www.ephpp.ca/Tools.html.

Gerschick, T.J., & Miller, A.S. (1995). Coming to terms. Masculinity and physical disability. In D. Sabo, & D.F. Gordon (Eds.), *Men's health and illness gender, power, and the body*. Thousand Oaks: Sage.

Gibbs, L. (2005). Applications of masculinity theories in a chronic illness context. *International Journal of Men's Health, 4*(3), 287-300.

Gibbs, L. (2007). Identifying work as a barrier to men's access to arthritis self management programs. *International Journal of Men's Health, 6*(2), 143-155.

Gibbs, L. (2008). Men and chronic arthritis: Does age make men more likely to seek self management services? *Generations, 32*(1).

Gibbs, L., & Oliffe, J. (2004). Reaching out to men: A social marketing approach. In R. Moody, & A. Hulme (Eds.), *Hands on health promotion* (pp. 356-363). Melbourne: IP Communications.

Gibbs, L., & Reidpath, D. (2005). What about the men? Gender differences in utilisation of chronic illness self management services. *Australian Journal of Primary Health, 11*(1), 78-81.

Gibbs, L., Kealy, M., Welch, N., Daly, J., Green, J., Small, R., et al. (2007). What have sampling and data collection got to do with good qualitative research? *Australian and New Zealand Journal of Public Health, 31*(6), 540-544.

Green, J., Willis, K., Hughes, E., Small, R., Welch, N., Gibbs, L., et al. (2007). Generating best evidence from qualitative research: The role of data analysis. *Australian and New Zealand Journal of Public Health, 31*(6), 545-550.

Hammersley, M. (1995). Theory and evidence in qualitaitve research. *Quality and Quantity, 29*, 55-66.

Lincoln, Y.S., & Guba, E.G. (1985). *Naturalistic inquiry*. Beverly Hills, CA: Sage.

Moher, D., Jadad, A.R., Nicholl, G., Penman, M., Tugwell, P., & Walsh, S. (1995). Assessing the quality of randomized controlled trials: An annotated bibliography of scales and checklists. *Controlled Clinical Trials, 16*, 62-73.

Moher, D., Jadad, A.R., & Tugwell, P. (1996). Assessing the quality of randomized controlled trials: Current issues and future directions. *International Journal of Technology Assessment in Health Care, 12*, 195-208.

Oliver, S., Harden, A., Rees, R., Shepherd, J., Brunton, G., Garcia, J., and Oakley, A. (2005). An emerging framework for including different types of evidence in systematic reviews for public policy. *Evaluation, 11*(4), 428-446.

Patton, M.Q. (2002). *Qualitative research and evaluation methods*, 3rd ed. London: Sage Publications.

Pinnock, C., O'Brien, B., & Marshall, V. (1998). Older men's concerns about their urological health: A qualitative study. *Australian and New Zealand Journal of Public Health, 22*(3), 368-373.

Public Health Resource Unit. (2006). 10 questions to help you make sense of qualitative research Critical Appraisal Skills Programme (CASP): Making sense of evidence, England.
Schofield, J.W. (1990). Increasing the generalizability of qualitative research. In E.W. Eisner, & A. Peshkin (Eds.), *Qualitative inquiry in education* (pp. 201-232). Teacher's College Press.
Spencer, L., Ritchie, J., Lewis, J., & Dillon, L. (2003, August). Quality in qualitative evaluation: A framework for assessing research evidence: Government Chief Social Researcher's Office.
Watson, J. (2000). *Male bodies: Health, culture and identity*. Buckingham: Open University Press.

Lisa Gibbs
Jack Brockhoff Child Health and Wellbeing Program
McCaughey Centre, School of Population Health
University of Melbourne, Australia

NORAH HOSKEN

DEVELOPING AND ENACTING AN ETHICAL FRAMEWORK AND METHOD FOR CROSS-CULTURAL RESEARCH[i]

INTRODUCTION: SEARCHING FOR AN ETHICAL FRAMEWORK

Cautionary 'anticipatory' ethical consideration and practice is essential in developing and implementing a research project involving ethnography and, more so, auto-ethnography (Tolich, 2010, p. 1600), particularly one that includes cross-cultural elements. Anticipatory means that ethnographers and auto-ethnographers should, among other procedures, usually gain informed consent before gathering and publishing data that can affect themselves and other people. This chapter reports on the continuing development of an ethical framework that underpins a current qualitative research project, *Searching for Recognition and Social Justice in Tertiary Education.* The project aims to place at its lived heart daily, enacted and active respect for the different and similar world views, culture and standpoints of its three main informants. As a written form is being utilised to communicate ideas about ethical frameworks, it seems appropriate to demonstrate ethical writing practice. Here, an illustrative tension becomes apparent in how to acknowledge the various contributions to the creation of this written chapter. I suggest this tension relates, at least in part, to differences between Western and African ontology (being), epistemology (knowing) and axiology (doing) (Bangura, 2011; Moreton-Robinson & Walter, 2009).

In what follows, I provide a brief summary of the current research project, describe the key ethical issues in writing and authoring this chapter, and go on to outline some important threads of the developing ethical framework. Other dimensions considered in the ethical framework are not addressed here due to word limit constraints. As the research project uses methods of ethnography, auto-ethnography and narrative, a reflexive style of writing is used. Tensions are explored; however, a conclusion is not drawn regarding questions of authorship and some other ethical aspects.

CONTEXTUAL SUMMARY OF RESEARCH PROJECT

The institutional ethnography explores the links between one Australian regional university's gendered, racialised and classed practices and its governance of social justice in higher education within the widening participation agenda. The analysis uses the narratives of three people from similar and different social locations, who would not 'normally' have attended a university. The exploration is viewed and

M. Vicars et al. (eds.), Discourse, Power, and Resistance Down Under, 27–39.

grounded from the interactive and continually constructed standpoints of an Anglo Australian female working-class background university social work academic staff member and PhD student (me, Norah Hosken), a South Sudanese Australian female Bachelor of Social Work student (Chagai Gum Malong) and an Anglo Australian working-class background social work faculty Professor (Bob Pease) who supervises me and the research project. My hope is that by promoting a complex, broader understanding and acknowledgment of the diverse, at times, competing interests and views about education and social justice the likelihood of preserving, advancing, respecting and aligning its necessarily different versions and progressions may increase. This research project uses Dorothy Smith's institutional ethnography (2005) as a framing methodology that is supplemented with autoethnography, cross-cultural mutual ethnography, narrative, Western feminist, African feminist and African research methods. This combination of methodology, methods, philosophies and ethics were developed in response to this particular research project's emerging contexts. In contrast to most institutional ethnographies I place more emphasis on understanding the complexity of the pre-conditions that shape and influence how the informants then interact at their site of study and work.

Chagai was a student in the TAFE welfare course where I was a teacher, and is a current student in the social work course where I now teach at a regional University. Chagai and I are student and teacher, friends, mutual mentors and co-founding members of the Geelong and South Sudan (Agouk) Community Coming Together Group. This research project we are both involved in represents a process and product that utilises parts of our separate and entwined life stories, experiences, goals and relationship.

BACKGROUND TO THE PROJECT

The Australian Human Rights Commission (AHRC) identifies education as one of five main factors affecting settlement of African Australians into the Australian community (2009, p. 5). As I have written elsewhere, 'social work and welfare work in Australia are placed in important and multi-dimensioned positions in relation to complicities, responsibilities and potentialities with this educational human rights issue' (Hosken, 2010, p. 1). There is little information about the knowledge, expectations and experiences of South Sudanese Australian women (Hatoss & Huijser, 2010) and working-class background women, in Australian tertiary education.

Chagai and I had many conversations about education:

> When I finish these courses I will be the first Sudanese woman to study and complete the Diploma of Community Welfare Work and the Bachelor of Social Work in Geelong. I will set an example to the women in my community and my children because the environment and culture we grew up in taught us that women should stay at home and we are lucky to be here in Australia at all.

> This is what led me to choose this path to work with vulnerable people, to understand their problems and let them feel that we feel for their feelings; and show them that we are fair with them. I want to continue to learn to help people who have problems that are difficult to discuss because of language, sickness, culture and gender. Sometimes education and knowledge can make a small or big change (Chagai Gum Malong, Research Interview Transcript 8.09.11, read verbatim from an email 20.09.10).

In keeping with one of the key principles of Smith's institutional ethnography, as a feminist methodology (1974, 1988, 1990a, 1990b; 2005; 2006), this project acknowledges people's lived experiences and knowledge as the entry and ongoing check-in (checking) points, for this study. Institutional ethnographies examine the ways in which people's everyday practices, talk and texts are embedded in a prevailing set of social or 'ruling' relations (Campbell & Gregor, 2004, p. 33).

The Hosken 2010 reference is 'my' first published article that drew from my experiences as a teacher at TAFE and University. I feel tension between pride at having published and discomfort in the knowledge that the Anglophone academic conventions and privileges affected me in not including as authors or contributors the students whose discussions and questions were integral to the article. The journal editor, peer reviewers and Bob also all contributed in different ways. I did get permission from the students and included this as a general statement in a footnote. I am not sure now if that was enough. I feel I should have made more effort to organise the permission seeking in person, rather than phone and email, where longer discussions may have enabled deeper consideration of authorship or contribution issues.

WHO ARE THE OWNERS OR AUTHORS OF IDEAS AND WORDS?

As a 'newish' PhD student and university academic, I am absorbing an understanding of how to technically utilise Western academic referencing conventions. However, central to developing and practising an ethical framework for this cross-cultural research project is a respectful curiosity that creates awareness of different cultural views about knowledge, intellectual property and what is valued. I have become actively conscious through discussions with Chagai and other people, that Western academic convention is based on notions of individual intellectual property ownership grounded in a certain historical, philosophical, political and economic context that are not necessarily shared or valued by people from other cultures. Chagai and I have discussed and puzzled about our contributions to this project, and to our individual and combined academic and community endeavours over the last four years. We are exploring in a connected, event and circumstance-based way, how we know what we know, how we decide what are good ideas, and if ideas/words belong to individuals or groups, or indeed to groups over time, and to those who are dead but whose ideas/words are still invoked and in this way felt as living. After writing this sentence, I automatically felt the need from my training and absorbing of Western

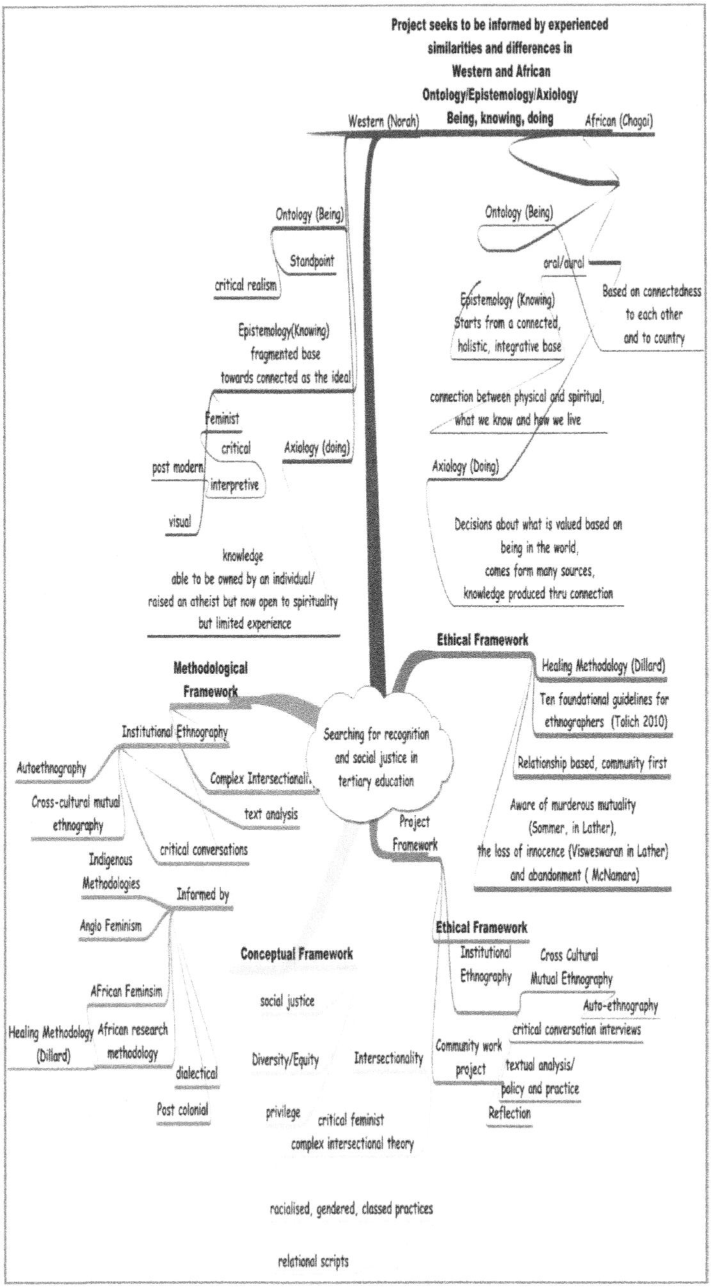

Figure 1. Project Mind Map-Searching for recognition and social justice in tertiary education

socialised academic behaviour about how to demonstrate being 'knowledgeable,' to somehow justify these ideas and words by referring to some 'authoritative written other.' Chagai did not feel this need at all, and wondered why it was not enough that she and I had experienced and said these words to each other. I still wrote the following sentence. This resonates with Pennycook's (1996) assertion that it is a historically developed (related to modernity, imperialism and Western humanism) and situated Western requirement that there are identifiable, preferably individual, author/owners of ideas, conversations and written words that need to be recognised and appropriately sourced (Pennycook, 1996, p. 201). As I have come to get to know Chagai (and myself) more in all the contexts and locations we relate to each other in, I am realising the extent of the impact of Western individualism and how the university I study and work in encourages this version of 'my-self.' When I talked with Chagai about why she did not seem taken with the university's academic referencing requirements she said there is no need to attribute words and ideas to one person in her community, and, it would not be possible or desirable, in most cases, to do so. It was not until later that year, in December 2011, when I was sitting under a huge tree with Chagai and many people from her remote village in Mayom Karraduit, Warrap State, South Sudan, listening to elders, village community representatives, and people speak and respond to each other, about their experiences, hopes and needs, that I felt the reality and importance of what Chagai had said. Stories, narratives are told from elders and family members, passed on and shared orally for many purposes, often to provide hope and guidance about important matters for communal pasts and futures, for living (Agouk Dinka Community, 2011). In this way I think I understood a little more that the South Sudanese narrative is not individual but communal, and the importance of this for necessarily intertwined survival, sociality and ethical living.

This chapter, then, utilises knowledge and ideas from a range of sources. I am the initiating researcher for the doctoral research project on which this chapter is based and physically 'compiled/wrote' with a flexible emerging outline in mind for its shape and content. The issue explored here of the impact of Western academic conventions on recognising and respecting different types of knowledges and contributions, has arisen in the project through ethnographic exploration sensitive to seeing similarities and differences in informants' lives and their connections to education and the university. Bob, in his role as my PhD supervisor, provided some comments on this chapter that I incorporated. I have not felt the need to include his contribution as authorial as (he did not seek it and) I do not consider his contribution to this chapter as major. (On reflection, this is likely to be an unfair or inaccurate assessment given how much I absorb from Bob's encouragement and knowledge as my PhD supervisor.) In this area of the project, however, I felt that Chagai's contribution, as key research informant and advisor was crucial to the content and enabled analysis for this piece of writing. It would not have been possible to write this chapter without the verbal comments; cross-cultural mutual ethnographic research discussions; and daily or weekly opportunities for discussing and enacting cross-cultural mutual understanding and pedagogy in both the research project and community project Chagai and I are involved in. The written words Chagai provided are contained to the one email referred to later in this article. Chagai read and approved of this chapter before submission. Chagai and I discussed her being a co-author, a contributor or being acknowledged in some other way in this and other future publications and presentations. My initial thoughts for this chapter were that Chagai be offered to be a co-author.

I consulted with Bob, who thought after reading an earlier draft of this chapter that perhaps I should consider the reasons and implications of giving Chagai's contributions as co-author status. A key concern Bob asked me to contemplate was if I would be comfortable with Chagai sharing the same responsibility I have for the chapter with potential criticism for any content inadequacies. He also raised a question about how including Chagai as a co-author might be interpreted from a variety of readers. Do I want to include Chagai in an attempt to afford my writing more legitimacy than if I am sole author? Does it protect me from criticism from some quarters? Bob raised these issues in the context of discussing the debates and controversy surrounding the co-authored Bell and Nelson article (1989). I immediately felt uncomfortable. Bob also asked if I had thought about how I would reconcile having joint authorship on this article but the requirement that my PhD thesis be as a single author. This made me feel unsure. Bob and I finished our discussion having left the issues unresolved. I sought guidance from my university's policy to clarify in my own mind what defines 'author.' The Deakin University Guide does not itself define 'author' but refers staff to a definition in the Australian Code for the Responsible Conduct of Research (2007).

> To be named as an author, a researcher must have made a substantial scholarly contribution to the work and be able to take responsibility for at least that part of the work they contributed. Attribution of authorship depends to some extent on the discipline, but in all cases, authorship must be based on substantial contributions in a combination of: conception and design of the project analysis and interpretation of research data drafting significant parts of the work or critically revising it so as to contribute to the interpretation. (NHMRC/ARC/Universities Australia, 2007, p. 5.1)

Although this definition seemed, initially, straightforward, it did not resolve for me the question of the status of Chagai's contribution. I think it is important to reveal and describe my process that is 'compiled/wrote' to further illustrate the tension and ethics in consideration of the questions, if it can be easily decided who authors/ writes/ owns, and what is a 'substantial contribution' in the required combination outlined in the above quote.

In writing this chapter, I read extensively, thought about, summarised and synthesised these different authors' ideas and words in the context of my lived experience and contemplations. I reflected on and reread the narratives and other materials from the research project from which this chapter draws. I discussed ideas with colleagues, friends and family, thought about many people's conversations, and typed these somehow combined words into my computer at different times of the day and night. Sometimes my subconscious mind sifted through ideas and events I had read or experienced to reveal of 'its' own accord some pattern or problem. I find that at times when I am in the shower, or walking, a thought or idea not linked to anything I was consciously thinking of, 'springs' into my mind. Also, as I typed at my computer, and paused hundreds of times to think, I summonsed and, at times, unintentionally, thought of, and then viewed in my mind, like a movie with sound, a collage of conversations that Chagai and I have had over the past four years. In this way, although Chagai has not written many words for this chapter, I believe numerous ideas, words and phrases in this chapter 'belong,' in the Western sense, to her. Also, some of, or senses and compilations of, the words, conversations, ideas, phrases of many other people (academics, family, friends, activists, people alive and dead) were not necessarily 'consciously summonsed,' but are also in this chapter and are not formally academically referenced. These are examples, I think, of what Pennycook refers to, when he asks "at what point does a phrase or an idea become owned? And at what point does it become public?" (Pennycook, 1996, p. 204).

The point of describing my writing process is to raise questions, or some doubt, about how other Western academics might write, to shake confidence in the assertion that it is always possible or easy to identify who owns knowledge, even in the Western academic convention that requires and prizes such identification. In this way, Tolich's reflection on the trouble with self (in auto-ethnography), although used in a slightly different context, seems an accurate and evocative description, "The word *auto* is a misnomer. The self might be the focus of research, but the self is porous, leaking to the other without due ethical consideration" (Tolich, 2010, p. 1608). In my writing process, it seems, the other also leaks to my 'self' without due ethical consideration.

I discussed authorship and acknowledgment status about this chapter in its current form with Chagai in person. Chagai read the chapter and approved the use of her narratives, the references to our experiences in South Sudan; and the chapter as a whole. Chagai did not seek authorship status and expressed satisfaction with acknowledgment of her contribution.

THE ETHICAL FRAMEWORK AND METHOD

The need for a considered ethical framework is great given the potential harms of research, and perhaps more so, of ethnographic and auto-ethnographic research and, where it involves researcher/participant relationships that cross cultures.

> Like an inked tattoo, posting an autoethnography to a Web site or making it part of curriculum vitae, the marking is permanent. There are no future skin grafts for autoethnographic PhDs. (Tolich 2010, p. 1605)

Tolich (2010) criticises some of the iconic autoethnographers for not practising the ethics and ethical processes they espouse in regards to anticipatory ethics, informed prospective consent and ongoing process consent. I remain concerned and mindful, trying not to be responsible for the creation of a process that could resonate with the introduction to Linda Tuhiwai Smith's book *Decolonizing Methodologies,* "Research" is probably one of the dirtiest words in the indigenous world's vocabulary" (2005, p. 1). Tolich offers ten guidelines for autoethno-graphers drawn from a range of sources, which, in my view, are unique, in this research genre's literature and ethical guidance, for the practical comprehensive clarity they provide. The advice or guidelines that include the following raise critical issues Chagai and I discussed in relation to our research project, before finding Tolich's article:

> Autoethnographers should not publish anything they would not show the persons mentioned in the text (Medford, 2006) ... Assume all people mentioned in the text will read it one day (see Ellis, 1995a). (Tolich, 2010, pp. 11-12)

I think the reasons Chagai and I discussed these issues without having such a specific guideline to provide (me as initiating researcher) an alert to do so, was due to the overall consideration for developing cross-culturally appropriate ethics, and lessons from my 30-year social work practice background and life experience. As a practising social worker, I often felt managers and governments could make policy decisions that hurt clients because they themselves did not have to directly witness, talk to, be in the same room with, the affected people. In 2011, Chagai was asked at a public forum what she might do to change politicians' minds about their policies regarding asylum seekers. I believe Chagai's response is based on a similar view to mine regarding the potential dangers of decision makers being too removed from the lives they affect:

> I told the people at the forum that these politicians should live for 24 hours with no electricity, no water and no food, and then consider if they too would take their family on a boat to somewhere in search of a better life ... the politicians should have men with guns torture or kill members of their family and see how far they will then go to protect their children. (Chagai Gum Malong, Research Interview Transcript, Verbatim 18.10.11)

Early in the project, Chagai and I discussed the 'Healing Methodology,' theorised and described by Cynthia Dillard (2008, p. 286) as part of an approach with essential ethics and methodology for working with indigenous and African women. This encompasses five principles of: 'unconditional love, compassion, reciprocity, ritual and gratitude.' These principles, also referred to as 'methodologies of the spirit' (Dillard, 2008, p. 287) are proposed as a way to 'honour indigenous African cultural and knowledge production ....' According to Dillard (2008, p. 287), love as practiced through 'looking and listening deeply' is an experience that creates 'more reciprocal sites of inquiry ... so that we know what to do and what not to do in order to serve others in the process of research.' Dillard's second principle, compassion, as a methodology requires the researchers to 'relieve communities of their suffering through the process of activist research.' It means the researchers must have serious and ongoing concerns for the research participants and want to bring benefits to them through their research. Within the third principle of healing methodology, seeking reciprocity, the researchers must have their 'intention and capacity to see human beings as equal, shedding all discrimination and prejudice and removing the boundaries between ourselves and others' (Dillard, 2008, p. 288). If the researchers continue to perceive themselves as 'researchers and the other as the researched,' or if they continue to see their own research agenda as more crucial than the needs and concerns of the research participants, they – cannot be in loving, compassionate, or reciprocal relationships with others (Dillard, 2008, p. 288). In the fourth principle of ritual, Dillard (2008, p. 289) states that as researchers 'we … are always one with spiritual reality, not removed from it as has been the ethos of Western research traditions.' Finally, in the fifth principle of gratitude, Dillard (2008, p. 289) describes the need to 'be thankful for the work of research as … a healing process for ourselves and others.'

In a research interview session, I read and summarised Dillard's principles to Chagai. Chagai then responded relating Dillard's principles to her own position in regard to the research:

> Oh my God, I love her. I do not know who she is but I love her principles. I like them all. Without the good relationship between you and me, I would not be participating in the research. Without having a connection, between you and me, and knowing how the connection came up, the connection of love, I would not be participating in the research. I would have said no I am busy. She [Dillard] is exactly African. (Gum Malong, 2011a , Interview Transcript Verbatim 15.08.11)
>
> I think important values in the research need to include:
> - Relationship has to be there before the research
> - Respect – I have to care about who is around me, have to care about my brother, my mum, my sister, my children -have to care about the whole community.
> - There is no I, the research will be judged by the whole community and should make my community proud.
> - Community first- it is others first before yourself. (Gum Malong, 2011b Chagai Gum Malong, Interview Transcript, not full verbatim, summarised, 15.08.11)

Chagai's affirmation and additions to Dillard's principles are a key inclusion in the development of this research project's ethical framework. The main reasons for placing importance on developing an ethical framework for this study include: wanting to actively respect research informants and the people/communities identified as being significant to them; preventing or minimising potential for exploiting the time, knowledge, goodwill and relationships of informants; to respect and learn from the similarities and differences between Western and African ontology (being), epistemology (knowing) and axiology (doing) (Bangura, 2011; Moreton-Robinson & Walter, 2009); and to engage in mutually beneficial and reciprocal research learning relationships.

MUTUAL MENTORING LED TO CROSS-CULTURAL MUTUAL ETHNOGRAPHY

Incorporating, supporting and in addition to, Dillard's (2008) and Chagai's (Gum Malong, 2011b) principles, the development and use of the adapted ethnographic research method, 'cross-cultural mutual ethnography' (Hosken, 2010) is also at the core of the ethical framework being developed in the current project. This method grew out of 'my' welfare and social work education and teaching practice and individual discussions with many students across similarities and differences, including what I came to think of and call, 'mutual mentoring' discussions with Chagai. These discussions often commenced from either the student or me identifying some additional or different explanation or support was needed to navigate the course or unit of study successfully. These discussions moved from being more one way – teacher to student – when I acknowledged how much I was learning from the ongoing interactions with students, particularly learning to teach in a way that saw, respected and included knowledges other than the dominant middle-class male Western knowledge that I think permeates me as an Anglo-Australian, and the institutions I teach in. The four-year association with Chagai includes being teacher/student, mutual mentors, community group colleagues and friends. A year or so before beginning the research project I started having conversations with Chagai about the idea of the project and her potential involvement. We talked many times about the importance we both place on education, the various barriers to achieving educational success and how to provide inclusive learning environments and teaching and assessment practices. The possibility of the PhD project being a step towards us writing a book together, or Chagai writing a book or other publications on her own, or with others was discussed. Chagai also mentioned the learning from her involvement in the PhD project helping in her own future studies.

After commencing the study, and searching the literature unsuccessfully for a method that incorporated the principles and process that made the mutual-mentoring work, I then developed the mutual mentoring into a method which I named 'cross-cultural mutual ethnography.' This method aims to provide a level of dynamic complexity and detail only made possible, I think, through ongoing in-depth discussions and analysis grounded in a relationship that involves a two-way exchange of information and knowledges, and the two-way experience of the vulnerabilities and growth inherent in that exchange. Another related reason for my own inclusion in the cross-cultural mutual ethnography was as a potential safety brake, a guard on disclosing too much, or not relevant, personal information. As I wrote in the Ethics Application:

> One of the reasons for including me in the research as a participant was to acknowledge some of the similarities between myself and the other proposed main participant, and that the risks involved, the potential for harm might also then inevitably involve harm for myself. Any potential harm resulting from the research would thus affect both the main potential participants (and perhaps the other two potential senior academic participants), notwith-

standing this may occur in different ways. (Hosken, Ethics Application, 2011)

The development and use of this method then lies at the core of the ethical framework of the project. Another key element in developing and enacting the ethical framework, encouraged by Anglo and African feminist frameworks, is the attempt to combine theory, practice and activism in the day to day life, informal and formal conversations and behaviours that comprise the relationships between the three participants (and others) enacted in the process of "doing" (Fenstermaker & West, 2002) class, gender, race and, research, in this study. The Geelong South Sudan (Agouk) Community Coming Together project that I am involved in with Chagai, initially sat alongside the study. The reasons and process of the community project becoming increasingly integrated into the study offers important insights to support one of this study's early claims that unions of people working with, across and at times (somewhat) through, differences has to be based on need and partial commonality (Jordan 1985 p. 47, in Friedman, 1995, p. 40) that begins with Mohanty's (1989) "assumption of the others agency" (in Friedman, 1995, pp. 15-16), rather than from what Allen describes as "white guilt ... (induced) ... generosity" (2004, p. 127). The research relationship and the community work relationship between Chagai and I inform and construct each other.

Other parts of the ethical framework include an open and ongoing discussion of types and levels of co-authorship, co-researching and shared decision making about the research questions, methods, analysis, write-up and different sorts of publications and actions. I continue to explore and discuss with Chagai and Bob strategies to identify and minimise the risks associated with what were classified in the Ethics Application form as an unequal power relationship, a dependent relationship, a relationship involving dual and triple relationships between me as the initiating and primary researcher and Chagai as the South Sudanese Australian student participant and advisor, and Bob as PhD supervisor, project participant and my work colleague. Due to word constraints here, I cannot share the discussions and thinking in the project that address important considerations around research, power, friendship and the minimising the potential for abuse and abandonment.

Chagai and I had several conversations about the advantages and disadvantages of anonymity, or using her real name, in this research study and eventual report, and in this chapter. Chagai chose to use her real name so she is recognised for her, and her community's, contributions. This was only decided though after we discussed measures for: determining what areas were suitable for discussion in the research; the use of continuing process consent; Chagai being provided with the raw transcripts of her narratives and then the sections chosen for inclusion in the study for her decision on inclusion and editing; input into any analysis; Chagai having final decision on what is included of the material she provides in the overall project; and Chagai's permission and opportunity to edit or remove her own narrative from any publications.

Finally, I suggest that the ethical framework development, necessary in its own right, also enhances the ability of the research process to explore and reveal some aspects of the nature and impact of 'Western-ness' on those who teach and study at the University, and on social justice.

NOTES

[i] Note: Chagai Gum Malong has made a significant contribution to this chapter in her roles as research informant and research project advisor. How this contribution should be recognised, and if it is actually authorial, is considered in this chapter.

REFERENCES

Agouk Dinka Community. (2011). Community meeting narrative under tree.

Allen, R. (2004). Whiteness and critical pedagogy. *Educational Philosophy and Theory, 36*(2), 121-136.

Australian Human Rights Commission. (2009). *African Australians: A report on human rights and social inclusion issues*. Sydney: Australian Human Rights Commission.

Bangura, A. (2011). *African-centered research methodologies:From ancient times to the present.* San Diego: Cognella.

Bell, D., & Nelson, T. (1989). Speaking about rape is everyone's business. *Women's Studies International Forum, 12*(4), 403-416.

Campbell, M., & Gregor, F. (2004). *Mapping social relations: A primer in doing institutional ethnography*. Walnut Creek, CA: AltaMira Press.

Dillard, C. (2008). When the ground is black, the ground is fertile: Exploring endarkened feminist epistemology and healing methodologies in the spirit. In N. Denzin, Y. Lincoln, & L. Smith (Eds.), Thousand Oaks, CA: Sage Publications.

Fenstermaker, S., & West, C. (2002). *Doing gender, doing difference: Inequality, power, and institutional change*. New York: Routledge.

Friedman, S. (1995). Beyond white and other: Relationality and narratives of race in feminist discourse. *Signs, 21*(1), 1-49.

Gum Malong, C. (2011a). Cross-cultural mutual ethnography narratives. South Sudan (Agouk) Dinka Community.

Gum Malong, C. (2011b). Discussion of ethical research principles for research involving African women.

Hatoss, A., & Huijser, H. (2010). Gendered barriers to educational opportunities: Resettlement of Sudanese refugees in Australia. *Gender and Education, 22*(2), 147-160.

Hosken, N. (2010). Social work and welfare education without discrimination. Are we there yet? *Practice Reflexions, 5*(1).

Moreton-Robinson, A., & Walter, M. (2009). Indigenous methodologies in social research. In M. Walter (Ed.), *Social Research Methods* (pp. 1-18). Melbourne, Australia: Oxford University Press.

NHMRC/ARC/Universities Australia. (2007). Australian code for the responsible conduct of research. National Health and Medical Research Council, the Australian Research Council and Universities Australia.

Pennycook, A. (1996). Borrowing Others' words: Text, ownership, memory, and plagiarism. *TESOL Quarterly*, 201-230.

Smith, D. (1974). Women's perspective as a radical critique of sociology. *Sociological Inquiry, 44*(1), 7-13.

Smith, D. (1988). *The everyday world as problematic: A feminist sociology*. Boston: Northeastern University Press.

Smith, D. (1990a). *The conceptual practices of power: A feminist sociology of knowledge*. Boston: Northeastern University Press.

Smith, D. (1990b). *Texts, facts, and femininity: Exploring the relations of ruling*. London: Routledge.

Smith, D. (2005). *Institutional ethnography: A sociology for people*. Walnut Grove, CA: AltaMira Press.
Smith, D. (2006). *Institutional ethnography as practice*: Rowman & Littlefield Pub Inc.
Smith, L. (2005). *Decolonizing methodologies: Research and indigenous peoples.* Zed.
Tolich, M. (2010). A critique of current practice: Ten foundational guidelines for autoethnographers. *Qualitative Health Research, 20*(12), 1599-1610.

Norah Hosken
Social Work, Health & Social Development
Deakin University, Australia

ANGELO BENOZZO, NEIL CAREY, TARQUAM MCKENNA AND MARK VICARS

GAY AND QUEER COMING OUT INTO EUROPE (PART 1)

Fragmentary tales of crossing national/cultural borders

INTRODUCTION

When South Pacific opened in 1949 racial segregation was *the law* in the US southern states. Lawmakers in the south tried to ban the following song ... but failed. It was a monumental act of courage by the writers and producers of South Pacific to refuse to back down. In this chapter we will present our ideas around how research practice can helps to query the boundaries between what is considered "professional" and legitimate knowledge and behaviour and what is not as four 'queers' in academe. Academic culture when thought of as primarily being about the transmission of knowledge situates the roles of gay men who are researchers in specific ways of being, belonging, acting and speaking.

YOU'VE GOT TO BE CAREFULLY TAUGHT

You've got to be taught to hate and fear
You've got to be taught from year to year
It's got to be drummed in your dear little ear
You've got to be carefully taught
You've got to be taught before it's too late
Before you are six or seven or eight
To hate all the people your relatives hate
You've got to be carefully taught
Careful the things you say,
Children will listen.
Careful the things you do,
Children will see.
And learn.
Children may not obey,
But children will listen.
Children will look to you
For which way to turn,
To learn what to be.
Careful before you say,
"Listen to me."

M. Vicars et al. (eds.), Discourse, Power, and Resistance Down Under, 41–57.

> Children will listen.
> How can you say to a child who's in flight,
> Don't slip away and I won't hold so tight?
> What can you say that no matter how slight won't be misunderstood?
> What do you leave to your child when you're dead
> Only what ever you put in its head
> Things that your mother and father had said
> Which were left to them too.
> Careful what you say, children will listen
> Careful you do it too, children will see and learn. Oh!
> Guide them but step away,
> Children will glisten.
> Tamper with what is true
> And children will turn,
> If just to be free.
> Careful before you say,
> "Listen to me."
> Children will listen ...
> Children will listen!
> Children, children will listen

As the song says when you "tamper with the truth people will turn" and this chapter is a series of evocations to the readers to reflect on how the categories of professional and personal lives have come to be disrupted by the interaction of our selves with our queer epistemologies. We have to be careful before we say 'listen to us' – You too have all 'got to be taught before it's too late,' 'You've got to be carefully taught.'

This chapter is very performative – it reveals in our speaking, acting and being in life from a "queer" perspective. We are doing this because we want to provide a possibility for all those involved in research teaching and learning interface re-rite themselves as they consider what we are saying and performing. In many senses the symposium that we performed at the AQR/DPR Downunder event was a challenge and positioned us a queers as a basis for questioning existing truths. Grace and Hill (2001, p. 4) suggest that:

> Queer knowledges can proffer a location where identities grow. They enable learners to challenge heterosexualising discourses and heteronormative ways. ... In doing so they situate queer performance as an alternative pedagogy that often forms new directions for personal development ...

Research as a way of knowing and being can help textualise the spaces of the gay man as an academic and his sense of belonging but it also interferes with and contests dominant pedagogies understandings; especially for those having heterogendered sexuality.

Working in academe Mark, Angelo, Neil and Tarquam all believe and hold to the notion that the individual's identity is constituted by processes originating within the *cultural environment and its institutions*. Mark and Tarquam work in a Victoria University School of Education. And they hold to the notion that Schools of Education are ordinarily by definition conformist and follow a "middle road" around societal change. Neil and Angelo work in Manchester, UK and University of Valle d'Aosta, Italy respectively where they are researchers and psychologists in psychology and health. Teaching, education and psychology are all professions that lag behind law as two of the more conservative professions. As the eminent scholar Peter McLaren (1999) notes, teachers cannot avoid participating in those performances and actions that he calls rituals.

This chapter sets out to use narrative and a 'play' to demonstrate how our collective and individual senses of queer intersect, coalesce and separate. We are proposing that there are rituals which contribute to our own experience of marginality, privilege or visibility. The chapter also sets out to critique how we are accommodated and embody the vast array of differences in the context of our schools and the academy especially as we research queer knowing. The chapter will contend that there is still a way to go as universities are essentially heterocentric and their Research Centres don't include us and render us invisible as they often ask, "What is the issue here?"

You will uncover how we as queer researchers react to our inclusion and exclusion and what roles the academy puts upon us to 'enact' through their notions of diversity and how our differences are at the core of our individual and collective selves. We will ask how can our resistant roles possibly to be seen?

The work we have been doing as educators and psychologists is founded on a belief in a theory of enactment of self; based on many years of experience as a researchers-as-teachers and teachers-as researchers – we are educators, counsellors, psychologists and psychotherapists. Our research and presentation in scholarly articles, in journals and writing and in "just being real" encourages the expression, rather than the *repression* of our queer difference. But is this at a cost? Our work in the universities we have worked in sets out to assemble and re-assemble and explore how queer stuff troubles some and is blithely ignored by others. It is that we note that we are framed primarily by the eyes of others and are subjected to their normalising gaze. This is also what Dubois (1973) calls the double consciousness of looking at oneself through the eyes of the other. Ultimately the symposium that this chapter is based on was a performance that asked you as the audience (and reader) "What happens if we use research and research around performativity of identity as a way of knowing and inquiry to serve to describe and highlight the *denial of identity* and *ethical care* of self for queers, gays in our societies and especially for us in our schools?"

We belief the self is constructed through daily encounters with a variety of "roles," and we deliberately use this term in the 'psychodramatic' sense. Role refers to the functioning form the individual assumes in the specific moment he reacts to the specific situation in which other persons or objects are involved (Marineau, 1989). The performance of our stories is based on the belief that sense of self (psyche) is constructed and contested in the lived action of *encounter* and *engagement*. As queer researchers we are writing around what transgresses and traverses boundaries: but it is the space for encounter and engagement that we are most interested in and especially how we might as researchers undertake research that brings in to the arena evidence of how our schools and our societies are still places of silencing, making queers invisible victims of gross and subtle discrimination.

The subtle discrimination cannot be fully be explored this chapter but returning to Moreno we are reminded that role evolves directly from the individual's interaction with the world, *even before the person has a unifying sense of "self"*. Hence the opening song. For the young women and men in schools and universities we work in and research indicates that "...role playing is prior to the emergence of the self. Roles do not emerge from the self, but the self may emerge from the roles" (Moreno, 1964, p ii, 157). The academy, the Schools of Education, the Faculties of Education that we work in and engage in daily can and do offer evidence of and indicate the extent and validity of the enactment and portrayal of stereotypes of queer, gay and lesbian people.

Our research and performative enactments display discourses of different sexualities are made invisible under hetero-hegemonic discourse, unless participants are allowed to express their own epistemologies of their closets. *Silences*, Sedgwick (1990, p. 3) contends in *The Epistemology of the Closet*, surround the closet, not just one silence, but also several "that accrue particularity by fits and starts, in relation to the discourse that surrounds and differentially constitutes it." For queer researchers such as us the evocative ethnographies (Ellis, 1997) are ways of knowing and becomes an exploration, which for students and teachers are like a breaking of silences and a memoir of the queer emergent life stories.

Despite research indicating that sexual culture is active in the shaping of school relations, Ken Plummer's (1992) finding that sexuality and particularly queer sexualities has been absent from curriculum theorists' and policy makers' thinking still hold to Sedgwick's (1990, p. 3) contention that many silences surround thinking about queer sexualities and the gender defining processes but how are these researched in our experiences? Health and education, notions of difference and appreciation of emerging understanding of alienation and enforced silencing in learning are all central to what we can see as 'agitations' that we can still see especially in our schooling systems right now.

These research agitations should set out to redress rigidities and inequalities centred on the binaries of the sex/gender system: especially the femininity/ masculinity and homo-hetero divides. Johnson and others illustrate that this dynamic interplay between New Right movements and sexual radicalism has four discursive clusters: those aligned with Neo-Conservatism, 'the family' and reactionary homophobia; neo-liberal attitudes towards sexuality and privacy; a mainstream social fabric which supports sex education; and an emergent and radical sexual theory and practice.

The combination of the above elements is critical in that they generate qualities such as "the incitement/containment of heterosexual desire, the freezing of gender roles, and the stress on biological and social reproduction" (Johnson 1996). Tierney and Lincoln (1997) constantly emphasize the use of narratives as a form of social inquiry. Our evocative queer ethnographies are not new but do serve to give space to the queer lesbian and gay self-defining people in academe.

Savin-Baden and Fisher (1992) use criteria of plausibility and trustworthiness rather than reliability and validity to examine the honesty of ethnographic data we gather. The trustworthiness of the research spaces we engage in and encounter queer others drives this inquiry – and the plausibility of being who we really are is at the core and for this way of researching. The principle of empathic engagement in the encounter with another being is at the core of the work. Ellis (1999) notes that we "... come to understand yourself in deeper ways. And with understanding yourself comes understanding others. Autoethnography provides an avenue for doing something meaningful for yourself and the world" (p. 672). The queer space and the research practice there emphasises that ethnographies are community-focussed processes. Addressing interactions and the dynamic of human activity and Ellis especially remind us that as researchers we are moreover communicating humans who are 'studying' other who are humans and we are inside what we are studying.

The research we are undertaking around queer stories are evocative narratives where biography and fact serve to fracture the prevailing boundaries that normally are not considered in the world of these people. Storytelling, theatre and other artful practices are for us as researchers the way in which the boundaries of self and others are re-aligned.

These models of praxis are emerging as are the findings – our emergent ideas around inquiry or research with disenfranchised people is complex and contested. The research we are examining has a quality of provocation. Remembering Ellis, Savin-Baden and Fisher have all called us to recognize the multiplicity of ways of knowing and to evoke in us new ways of understanding our truths. The evocative autoethnographies as performed here, as stories, have at their core the self, as noted in the model of research described earlier. It is their flesh and blood emotions that will compel people to listen and hear the findings of these stories. Our research as evocative ethnography is dependent and on being researchers ourselves and doing research from our singularity and collective queer perspectives. The enactment of multiplicity as opposed to singularity of our selves is called for. It is this that becomes the weave in to which are threaded the community of straight and queer stories collaboratively. These stories are the intersubjective bridge that can respectfully loosen the ties that bind. These stories are the call to DEEP knowing and are founded on the lives of the women, men and others who we have not been heard.

MONOLOGUE – TARQUAM – EVICTION 1 – OFF THE ISLAND – INSTITUTIONAL CREATIVITY ACCOMMODATES OUTSIDERS' SEXUALITIES

In my primary and secondary schooling in Tasmania, there was no opportunity to be open and gay as a student. The one or two teachers that I had who were identifiable as 'outsiders' because they were artists and musicians; and who might have been gay, gave up the profession. It was only in my university training years from 1972 that any considered sense of the self being so much a "teacher" and a gay man, occurred .... I seem to recall that I was known as gay but there was not any place to talk about this and sexual expression was fairly much 'on the run.' In Melbourne one holiday I was at a beat. The poofter bashers were chasing gay men and the police were chasing the bashers and the gay men. It was all very scary in the seventies in Australia.

After I had been teaching for two years I was awarded a Rotary Foundation Educational Award – a very prestigious scholarship to study in the UK. I had hoped to study art therapy at Hertfordshire and return to Tasmania to work with school students using art therapy. The community of scholars in the special schools in which I worked had one collective voice. "How come he got one of those scholarships?" In retrospect I realise I responded to these questions out of a place of feeling a subtle and covert homophobic commentary.

The heterosexual teachers I worked with took for granted their daily capacity to represent themselves. They spoke of their social lives, the straight literature, straight television, and uncritically presuming dominance and "a lifestyle". My then partner never got mentioned openly. In this period of my life I was very heavily depressed. In a clinical context the depression was reactive, a reaction to an identity as a gay man being made invisible within the social and educative community. I taught and attempted to apply the training I had in drama and arts to the special educational context. Many colleagues spoke of me 'not being used' adequately in my teaching role. One colleague told me that the way I was 'treated' was wrong. She referred to the gossip, innuendo and invalidation of my ability as a very successful scholar who had worked with the world leader in drama education. I naively believed the community reaction in Hobart was some form of jealousy or the 'tall poppy syndrome.' And yet I had a nagging doubt that it was because I was gay.

THE PERFORMANCE

Gay and queer coming out into Europe: Fragmentary tales of crossing national/cultural borders

This fragmentary performance ethnography is per se a coming out story organized around 4 acts each of which is played in different places.

ACT 1 MANCHESTER – AN EXCITED COMMITMENT TO EXPLORATION: REPRESENTING A CONTRACT FOR THE RESEARCH.

ACT 2 VENICE – AN EMBODIED DREAMSCAPE OF SEEMING SAMENESS

INTERMISSION

ACT 3 TRIESTE – LANGUAGE FRAUGHT WITH ITS UNTRANSLATABLE BORDERS

ACT 4 MANCHESTER AND MILAN – GAY AND QUEER OBSERVATIONS: BODIES ON VIEW

PROLOGUE

What follows is a scripted performance. It is a 'playful' representation of the (real and imagined) meetings of two European gay men who begin to be aware that, as they share similarities, they also embody a great deal of difference in their glance to life, world and their (homo)sexual epistemologies. We perform some of the joys and struggles when communicating across the boundaries of our ideographic histories, languages, theoretical approaches, national cultures and (homo)sexualities. We have been inspired by the idea that in critical and queer qualitative research, data are not only words but also the images, sounds, smells, dreams and memories that are carried by those of us who are simultaneously the 'subjects' and 'objects' of research. As such the script presents a fragmentary performance ethnography (Holman Jones) constructed around a series of imagined and real dialogues that represent our meetings and distances. We offer them in order to provoke and disturb the audience, as well as our own understandings about the conditions of being 'out' at the beginning of the 21st century, and how those understandings can be known through research practice.

This performance in four acts by Neil – an Irish man – and Angelo – an Italian man – it can be "listened/heard" through different foci that are intertwined: these are stories of coming out, they are narratives of difference and similarity, and they are also a representation of the proximities and distances in communicating differences across borders in which the two men reside. Perhaps they also reflect the nature of situatedness for all researchers/researched.

Actors:
Neil Carey
Angelo Benozzo

ACT 1: MANCHESTER – AN EXCITED COMMITMENT TO EXPLORATION: REPRESENTING A CONTRACT FOR THE RESEARCH

ANGELO: We meet in Manchester
And we discover that we embody
different visions of homosexuality
We look, in a different and/or equal way, at men
that we cross along the streets
We have been performing coming out
We have been writing about it
and we realize that we have different ideas on how to write about it
We have different ideas on who is our audience

Image 1. Manchester Pride (c) David Murat (permission granted)

NEIL: Drawn in and through our similarities, I simultaneously become aware of some of our differences when viewing the pull and being of 'coming out'; I crave a more edgy and critical Queer lens. I recoil now from my initial impulse to reject your desire to explore 'coming out.' Where once he was a peripheral political buffoon, Berlusconi now crashes almost uninvited into my sphere. I begin to be a bit more sensitized to the political force and violence of his sexualized talk (e.g. Squires, 2010). Derek Duncan (2005) helps me to better frame your cultural context as he highlights some ways in which Reading and Writing Italian Homosexuality may be a 'case of possible difference,' distinguishing it from those cultural literacies that configure my own 'coming out' in a Liberal, urban British context. This new lens on coming/being out further unsettles my previous un/certainties.

Image 2. Manchester Pride (c) Tim Pickstone http://flickr.com/timpickstone (permission granted)

In seeing the possibility for alliance, I stop to reflect, to work towards harnessing the productive potential in our earliest clash, and welcome the opportunity to meet and talk again from within, as Jacob Wick (2011) describes it: an "intimate space of distant proximity."

Image 3. (c) David Murat (permission granted)

Image 4. Coronation St actors, Manchester Pride float by Tim Pickstone, flickr.com/timpickstone (permission granted)

ANGELO: In this work
I want to dialogue, perhaps turn upside-down,
With my story and with Neil's story
With my story of this male that feels attraction to other males
I want to legitimize my cogmotions
I want to write to investigate how my story
Is caged by discourses:
racialized,
rationalized
gendered,
able-bodied
Something really new for me.

And so we meet, exchange, and attempt to inscribe a way of sharing through and across our differences and similarities. I like the idea that we resist the metaphor of entering each other's space to understand – with its shades of colonization. Or that we resist translations of all that is fruitful and spiky about our difference into what Jacob Wick (2011) calls a "harmless, apolitical 'third way.'" Instead, I look forward to excavating what is productive from the clunk and clash of our difference as we also rest and giggle in the shadows of our similarity, recognising as does Miwon Kwon that, our "commonality [is] based not on identification but on difference ... [and that] the radical potential of the queer community lies in our very fluidity, our lack of a common history, a common myth, a common homeland: we are a community of others ... and we must – constantly re-imagine the ways in which we live in our bodies, with each other, in the world" (2006; cited in Wick, 2011).

ACT 2: VENICE – AN EMBODIED DREAMSCAPE OF SEEMING SAMENESS

Venice is sparkling. We are lucky: at the beginning of July the sunshine and the air are not quite as white hot as they could be. I drag a suitcase behind me and walk around with my partner. Bridges, churches, stairs, calle, canals and again Fondamenta, stairs, arcades, churches, Palazzi, Campi – the similar structures and devices are everywhere here. All these spaces are wrapped up in grey-green water. I feel so deeply Italian in looking at Venice; so deeply proud of this place.

La Serenissima is so fragile. Venice is a crystal that asks for care. Its glorious history echoes at every corner. Within every Palazzo the subtext is a brothel. La Serenissima struggles to preserve itself, and rouses my need to take care of it.

Sparkles and the dark side. A beggar, then another and another. With undone, ruined bodies, they bow. My neo-liberal earth quickens my pace. Pain and fault, I turn my eyes away. Venetian music echoes in the air. Embalmed, white clad mimes somehow withstand the sun. This ceaseless bed oppresses me. The smell of decomposed fish. And I run.

Image 5. "Gondolieri" available at:http://commons.wikimedia.org/wiki/Venezia

Finally we are at the hotel where we meet.

Indeed, Venice is resplendent in its watery beauty. Its potential for Disney-like simulacrum evident in the mass of other tourists pounding the well-worn narrow streets, all hoping to see that corner that has not been seen before – perhaps to capture and consume it for our own pleasure. The warm, bright sunshine blinds my backward glance to rainy Manchester, and revels in the glow of scantily-clad tourist bodies strutting along the canal ways.

You wait with me on the bridge as the rest of the group, much more familiar with the city, set a pace beyond my arthritic knees. We laugh about our slowness, our temptation to dawdle, knowing how it grates those setting the pace. It feels good to be still, to sit and breathe, to be in Venice at your invitation, to bask in the generosity of this Italian patience.

And of course, the temporary stasis allows me to indulge my usual scopophilic pastime. In this moment, much of Venice recedes to the background as I focus on the striped jersey of a gondolier leaning on a bollard, waiting for his next customer.

His pose foregrounds the package on offer: the stretched shine of his tight-fitting uniform; smooth, tanned skin so in contrast to my own hairy paleness; buff guns flexing arms across pectorals that assure the safety of any ride to come. The

Image 6. 'Gondolieri' available at: http://www.flickr.com/photos/wahlander/-

Image 7. 'Gondolier' available at: http://www.flickr.com/photos/josemanuelerre/

ribboned straw hat, wholly incongruent to this Anglicized eye seems, at first, foolish and risible. But atop that lean, dark latin face it only adds to the potential for fetish, heightening the allure to the exotic other. Querelle flickers frenziedly on the sheet of my hastily made internal cinema screen. Tom of Finland's serried troops seep and ooze in my mind. Potentially embarrassing bodily connections threaten, and trigger an abrupt withdrawal from my reverie. As I come back into Venice I catch you watching too. As you feel the intrusion of my gaze, our meeting eyes crease with laughter; the recognition that we had both journeyed in a seemingly similar canal ride with our gondolier.

We walk on the Grand Canal. Neil and I. Sun and light blue sky. Buildings at the back and in front of me: a museum in the open air. He sits on the parapet in front of us: the first gondolier. We walk and approach him. He sees us. We look at him. He asks: "Giro in gondola." "He is asking if we would like to have a gondola tour," I translate for Neil.

Image 8. 'Gondolier' available at: http://www.flickr.com/photos/42192683@N07/

NEIL: "A gondola tour? Yes, but not now."
However it will be more interesting, not in the Gondola but perhaps somewhere else. Him and me? All of us? We? Only Neil and him?
The second Gondolier:
He's waiting for costumers. He's sitting. Light waves move his shapes.
A straw hat and black trousers.
A shirt with blue and white horizontal rows hug his muscles.
Suntanned and marble profiles.
I say to Neil: 'Look at him there".
Penetrating gazes: black eyes.
His bandaged shape exude desire.

NEIL: "The Latin male".
"The maschio latino knows to be an object of desire".
I've never thought of a gondolier as the 'maschio latino.' I am not.

Image 9. 'Gondolier' available at: flickr.com/flydime

NEIL: "Fassbinder upsets and plays with such images of ambiguous masculinity. Do you remember it?"
"Yes I remember".
However when I saw it I was 20 and I did not look at it in this way. For me it was one of the many tales of male attractions I was looking for.

The third gondolier:
He is not alone. There is a group.
I am 22 and it is November.
I saw them on the other side of the Gran Canal. I approach, I observe. I fantasize.
They laugh, speak and smoke
Athletic shapes and hard round buttocks attract me.
I retreat. I can't desire
The thoughts are sinful
It is better to turn back
Lasciar perdere.
It is better to escape.

We must go on, catch up. The rest of the group, our partners, our separate (shared) worlds are already well along.

INTERMISSION

Signore e Signori, Madame e Messieurs, Ladies and Gentlements, C'è ora un intervallo di trenta minuti. Vi chiedo di spostarvi verso il bar e di non perdere l'occasione di dare un'occhiata alle splendide donne e ai meravigliosi uomini in giro. Il 'Bar del ridotto' è una vetrina grandiosa. È il posto ideale dove guardare e farsi notare, dove toccare e farsi toccare, dove esibirsi e vedere magnifiche esibizioni. E se alla fine, qualcosa dovesse andare storto, bene allora rilassatevi e prendetevi uno o due drink. Il terzo e quarto atto inizieranno tra trentacinque minuti

REFERENCES

See Part 2 of this chapter for all references and the second act of the performance.

Angelo Bennozo
University of Valle d'Aosta
Italy

Neil Carey
Manchester Metropolitan Universoty
Manchester, UK

Tarquam McKenna

Faculty Arts Education and Human Development
Victoria University
Melbourne, Australia

Mark Vicars
Faculty Arts Education and Human Development
Victoria University
Melbourne, Australia

ANGELO BENOZZO, NEIL CAREY, TARQUAM MCKENNA AND MARK VICARS

GAY AND QUEER COMING OUT INTO EUROPE (PART 2)

Fragmentary tales of crossing national/cultural border

THE PLAY CONTINUES

ACT 3: TRIESTE – LANGUAGE FRAUGHT WITH ITS UNTRANSLATABLE BORDERS

ANGELO: Ma non è possibile continuare così. Io non capisco niente. Perché scrivi in questo modo! Ma tu non ti metti nei miei panni! Pensa a quanto è frustrante per me continuare a interrompere, a non capire, a chiedere. A tradurre parola per parola e non capire nulla. E' frustrante!!! Forse è meglio che interrompiamo, che non continuiamo più a scrivere insieme![1]

NEIL: Ah! Although the Italian is impenetrable for me I can see and hear that you're not happy. Is this the Latin Machismo that we talked about in Venice? Is this your hot Southern blood that I see before me?

ANGELO: No, this was not the Latin Macho. Era più 'la maschia isterica.'[2]

NEIL: Hysteria? Tell me?

ANGELO: Since we started to work together, since I started to read your writings, I have had some troubles in understanding what you want to say. I am not a mother tongue.

NEIL: Mmm. Speaking and working across the italian/english language divide is, I know our biggest obstacle here. Or is this about my style of writing?

ANGELO: It is about both. Yesterday after reading your writing we tried to look deeply into it line by line. And at a certain point one sentence became completely obscure and impenetrable for me. Explanation after explanation, non sense after non sense, reading after re-reading. Your writing became a loop of non sense for me, I felt like an idiot wasting your time.

M. Vicars et al. (eds.), Discourse, Power, and Resistance Down Under, 59–76.

Image 10. Angelo Benozzo personal photograph, July 9, 2000

NEIL: I can see the frustration. And yet I like what and how I'm writing. I know that it is full of cultural references, and that it seems convoluted and unnecessarily complex. But I like it almost precisely because it eludes instant transparency. Whilst being aware of the power dynamics being played out in our attempts to translate, I'm also interested in how what I write defies easy translation, an easy comprehensibility. My intention is not necessarily to exclude but to stretch the possibilities of how I can and want to be understood as 'out.'

ANGELO: In this 'language point' there are a lot of dynamics including working across mother-tongues. In that power dynamic I fear that I am out of control, completely dependent on your expertise.

NEIL: But I don't see the dependency in only this way. I am also dependent on you working in English - for my sake. I have only one linguistic frame here. However, I think this 'crossing language' also involves a host of other complex and reciprocal recognitions of our differences.

ANGELO: Yes. I agree. And this 'language point' implicates our different conceptions of writing: what it means, what it is about and how it has to be according to our audience.

NEIL: But in imagining an audience I want to resist knowing already who that audience is. I toy with the idea that my writing might allow an openness of meaning for some of the different readers who may want to engage with it.

ANGELO: Does it mean that you want to be clever? I remember once we were in Manchester and you told me that you wanted to be 'clever.' I do not know exactly what 'clever' means for me. I only know that this word has an allure that is quite dangerous.

NEIL: No! this is not about just being 'clever.' I yearn to play with the form and style of language that I adopt in writing. I want to toy and tease in representing me. I want to work beyond those more traditional academic frameworks through which I have become.

Image 11. Angelo Benozzo, personal photograph, July 18, 2011

Image 12. Angelo Benozzo, personal photograph, July 18, 2011

ANGELO: However, I'm convinced that a text must be clear, transparent and comprehensible. This is my position.

NEIL: It's not that I want to make my writing incomprehensible. But I do want to resist instant transparency. Perhaps in Cate Watson's (2008) terms, I want to be in the moment of the Baroque; to be outside of the modernist mode of representation. I want to acknowledge that language is always literary, never some sort of mimetic conduit or mirror that describes and represents neutrally. I want to become out of and in language that is open, that suggests rather than reveals or makes transparent, that leaves space for and requires both the writer and the reader to make sense in their own relation to the text. I suppose I want to reflect my own resistance to knowing me, or to expect that I can be known. Maybe, like Mair (2010), I have the Queer desire to "... overthrow [the] heterosexual privilege and heteronormative assumptions which drive so many prescriptive life narratives across Western and non-Western cultures."

Image 13. Angelo Benozzo, personal photograph, July 18, 2011

ANGELO: And last but not least: Neil, why this need to recall all these 'Protective Saints' (Mair, Watson, Jesus, Mary the Mother of God, Saint Peter ...)?

NEIL: I think you definitely got me there! I'm clearly feeling the need to adopt them as warrant for what I want to say. It's ironic that I feel the need to invoke this litany of protection to voice me.

Perhaps my approach to writing will always be a threat to voicing me in a way with which I am happy, that allows me to communicate in the academy, and, that works in collaboration with others.

ANGELO: However don't you think there is a risk here? The risk is that all the attention to the language becomes an obsession and we reach an impasse and we remain stuck?

NEIL: I really want to avoid that impasse, But I know how the politics of language are also a politics of being. These difficulties of language are not going to go away.

ANGELO: So, we are always 'in the point and in the process.'

NEIL: Perhaps we just need to recognise that reality and (re)commit (Adams & Holman Jones, 2011) to what is productive within that?

ANGELO: Well, let's see how that difference plays out.

ACT 4 – MANCHESTER AND MILAN – GAY AND QUEER OBSERVATIONS: BODIES ON VIEW

ANGELO: I look out the window of my classroom on an ordinary day. I am in the third year of secondary school, I am thirteen. My classroom is on the second floor and it overlooks the playground of the school, which is surrounded by a grey concrete wall. It is an ordinary day; perhaps it is sunny. I glance at the playground on the left side of the wall and I see writing in block capitals. I read and re-read. And again I read and re-read this writing as if the signifier has no meaning. I move away from the window: these words have no meaning. And after a few seconds I am at the window again. The wall shouts: 'BENOZZO CULO.'[3] It is a huge and enormous writing big enough to be seen from every side of the school. My mother works in this school. What she will say about this? I do not know if she has read it.

Image 14. Angelo Benozzo, personal photograph, June 9, 2000

What can I do? How to wash away this stain, this accusation, this humiliation? Who did it? I withdraw and hide the suffering inside myself. I pretend nothing had happened and I rise above it. After several years I came back to see the wall. Time has erased the mark. The wall has been painted. The marks have melted away, but it has taken a long time for me to be able to be ironic about the episode.

After that, every now and then, what I was suppressing re-emerged. For example when, in front of my sisters' appreciation of the marvellous Paul Newman, I stayed silent even if I shared their same emotions and thoughts. Silent and alone, I desired the Village People. Until I was eighteen I gave little room to the 'Benozzo culo.' Then the attraction and the erection became disruptive. At the beginning I didn't know where that erection was located. I said to myself: "I have an erection for a man so I am not normal?!" With difficulty, and very slowly, I gave in to the desire and went with the fantasy. Perhaps I really was culo, gay, homosexual. Very slowly I gave energy and legitimisation to the imagination of an erotic life with a man. Perhaps I really was gay. What's wrong, strange, twisted or bizarre about that? Now I can shout that I am culo, gay, homosexual … I am … I am … I am cool!! Yes, I am cool!!

All around me I felt the eyes of friends and colleagues. The eyes were loaded with one question. What are they wondering or thinking? They are asking: "Perhaps Benozzo is gay?!" But what makes them ask this? Perhaps my face, my gentleness, my posture, my hand gesture, an inflection of the voice, something in my body. What is that something that I embody and that is definitely gay? At a certain point I decided to grow a beard to strengthen my face. And even now, even today, I wonder about those glances from friends and colleagues: "Did they realise that I am culo, gay, homosexual …?"

However nowadays I leave hints and clues. I have a gay flag on my bag and sometimes I do not hide my appreciation of a male celebrity. Always prudently and with caution, always without excess.

Image 15, Angelo Benozzo, personal photograph, June 9, 2000

I have to watch out because somebody might understand. But what? To understand 'Benozzo is culo.' Be careful because somebody could be there outing you. So it is better to disguise myself, as I did at a Gay Pride event, with a hat and black sunglasses. I didn't want to be recognised; I didn't want to display myself.

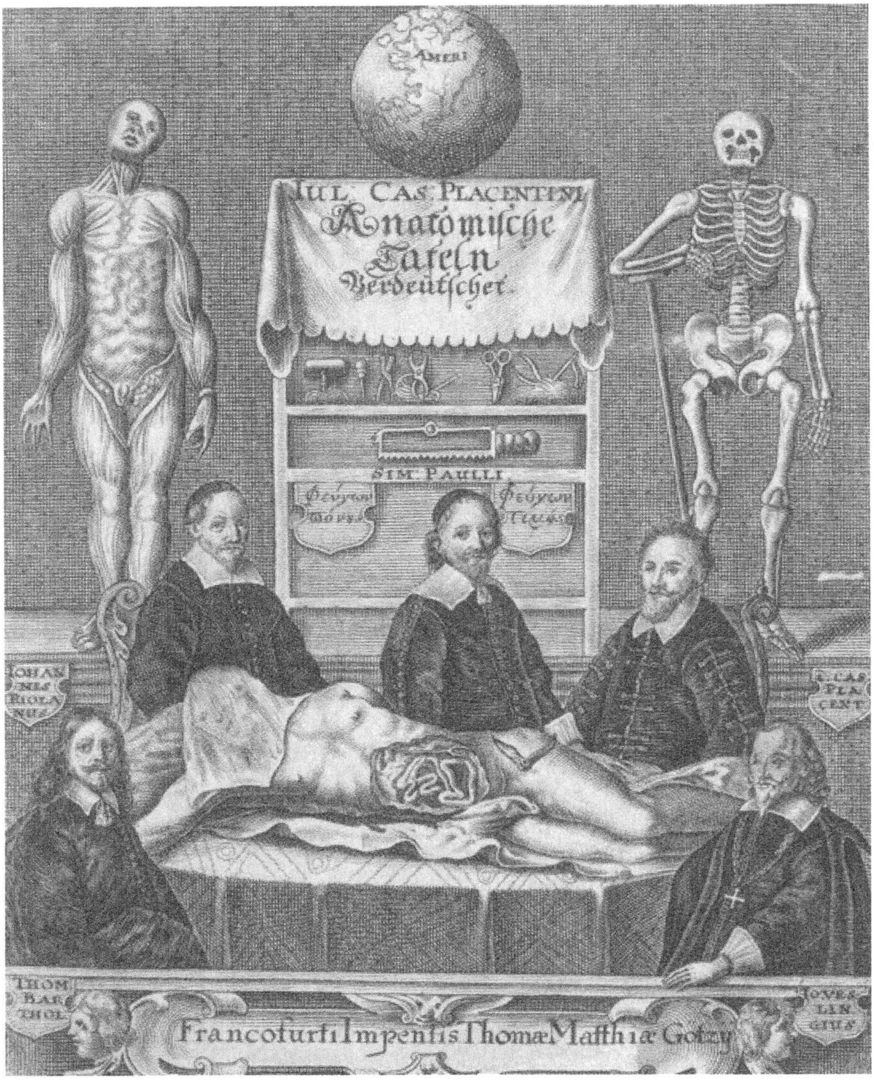

Image 16. Anatomische Tafeln ... Giulio Casserio (ca. 1552-1616) Frankfurt, 1656. Copperplate engraving (National Library of Medicine)

Splayed on the dissection table I resist the invocation to be screaming, in this display of knowing me and those like me. I look towards the corner and recognise the skeletal spectres of my parents already positioned as the progenitors of this queer specimen, as if some direct and phylogenetic causality might be detected.

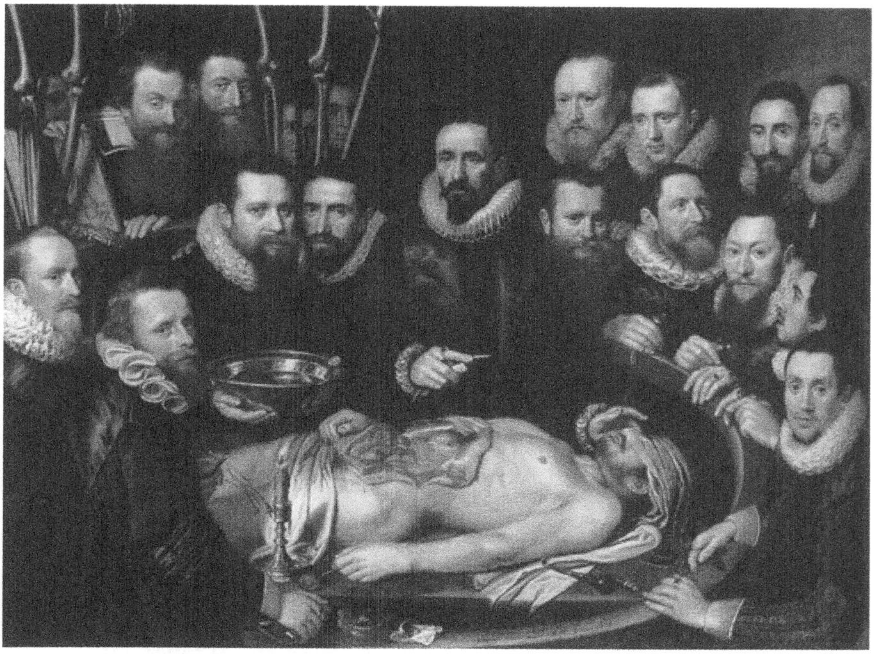

Image 17. 'Anatomy Lesson of Dr Willem van der Meer.'
Michiel Jansz Van Miereveld (1617)

I know that their nurturing capabilities have been thoroughly examined, questioned and measured as indicator, as predictor of my being out here – exposed. I carry these dead parents out with me in every calculated mince that I take; they are implicated in and implicate themselves in each attempted chic sashay performed by this body.

However, in failing to causally connect this queer body in its genetic and social genealogy, its individuality is poked and prodded, examined for its particularity. Surely there is some way of knowing who and what this body is. Its genetic coding is flashed on screen for all to see – are there peculiar loci here? Brain halves are compared for their symmetricallity; a hypothalamus is displayed in a Petri-dish, and representations of its anterior commissure and supra-chiasmatic nucleus are charted and calibrated against population norms. The musculature of this body's left hand and arm are measured against that of its right. And yet, none of this biology seems able to materially determine or pin me down.

Image 18. 'Dissecting the self' (Kings College London) available at: http://www.kcl.ac.uk/schools/humanities/hrc/life/dissecting/

This queer specimen stubbornly resists giving up any underlying cause to be known.

At the same time, I reflect that, being the object of such scrutiny from others' desiring to know, is nothing compared to the intense and complicated visibility to which I subject this body in my own conscious becoming. At times, guarded and overly self-aware, full of fear, of marking and being marked. Kissing and hand-holding, the lightness and knot of my scarf, the length of my hair, the limpness of my gait – all sites for regulation and surveillance that occupy space and linger in my head. At other times, I'm knowingly coy and joyfully projecting a determined visibility. Feigning nonchalance against the pillars of the dance-club or the canal-basin cruising ground, I negotiate the attentions of all those potential and imagined spectators making up the scope of my visibility. There are those who, I know, want to look; and those who I want to look away, from whom I want to hide my joys and pleasures. There are also those at whom I want to look, and of course those others to whom I want to be seen.

Such complex intra- and inter-viewing! As if all I had to do was to be a performer in control of and knowing my audience!

This body, ravaged and marked as it already is by other lashes of being should intentionally elude the disciplinary whip of knowing and being known. In this theatre of anatomy then, being screamer or out, being appropriate or appropriated is always only ever potential, possibilities to which I must acquiesce or resist, knowings not yet available into which I might instead relax.

Image 19. 'Rainbow flag' (c) David Murat (Permission granted)

EPILOGUE

And so our acts here, hopefully hint at some of the untranslatable knowings inherent in our commitment to perform research on and with coming or being 'out.' Two seemingly similar European men who harbour and nurture attractions for other men, mine some of their shared and distinct views on holding such desires. We hope the performance mirrors some semblance of the potential joys and difficulties in conducting qualitative social research: of being a researcher, of being researched, of making claims to know, and opposing some other forms of knowledge in relation to 'being/coming out.'

Monologue – MARK: For some gay men a queer space is often one that refuses to obey normative orders. A Queer space is that which engages with theatrical gesture; it can be overflowing with pure artifice, puns and playfulness and it can be a space where the everyday is distorted, distended and deformed. It is a space of re-appropriation. It is where the quotidian codes that govern gender and sexuality are misused and abused for 'perverse' purposes and pleasures. It is often in that in-between space where we begin to break through our skins and start to spin Other stories, some of which may be true, some may not. It can be a liberating space because it is affords sensuality in a world that frowns upon our artificial and excessive antics. Folding in the illusory, it is where a Queer aesthetic converges with social, cultural and political values and sensibilities. It offers a way of being in the world where that which has always been expressed unconsciously or unintentionally gets to be expressed consciously and intentionally. A Queer space is where we get to enjoy the transient artificial architecture of self, it where we embrace and acknowledge our queer lives as our own creations.

SILENCE can be the deadliest weapon and safest shield.
It is what I turned to and what I did when cornered by identities I never wanted to claim.
Retreating into **SILENCE** was a way to grin and bear it.
I heard my calling by age 7; they had a word for boys like me

SISSY!
PANSY!
POOF!
I now had names for something that I had hardly ever thought of at all.

At age 11, I graduated to a new knowledge
HOMO!
FREAK!
QUEER!

HATE
SHAME
REPRISAL.

At 15, I tried to suppress and disavow my homosexual feelings through exercise and attempted to construct a body that would encase my desire in a hard protective shell. I began refusing meals, disposing of food and deliberately starving myself. Seven months later I was recognised by my family as being anorexic and packed off to the doctor for diagnostic confirmation.

I understood how my life got discounted each day.
I felt
SHY
CONFUSED
ALONE.

At 16, I asked what's a HOMO? And the anger and hurt that had not been given voice

PREVAILED
ACCUMULATED.

At 16, having spent years listening to instructional voices, I was never quite sure if or how or when I should speak. Inside, I felt like my ship was sinking and that the primary concern of others was whether there should be fresh lemon slices in the finger bowls at dinner.

Named,
being spoken through a
language of:
MEASUREMENT
NORMS and DEVIATION
that
ACCOUNTED FOR, EXPLAINED:

PATHOLOGISED

I retreated into **SILENCE**.

My desire to tell stories has a shape. A chequered history, in living **SHAME** and pride.

When I was sixteen, I told my father I was Gay. He said, 'I know, you had better go and tell your mother.' As I entered the bedroom where she lay, her gnarled fingers misshapen through Rheumatoid Arthritis clutched the heated pad she used to alleviate painful joints. Heavy curtains were blocking out the afternoon sun, she had been prescribed new anti-inflammatory drugs that had aside effect of drying up her tear ducts. Twice daily a pipette containing synthetic 'natural tears' would lubricate her eye and keep them functioning.

I've something to tell you

I inwardly flinched,

I think I might be....

Why did I say 'think'? There was no doubt, in my body, I knew.

I haven't got a son I've got two daughters"......What have I done wrong? ...It is wrong; it's in the Bible....You'd better leave.

The silence between us irrevocably fractured, she turned her face away from where I stood and started to cry. The tears came quite freely now. I abandoned the melodrama that was being played out behind the net drapes of my parent's house and fled, not quite skirts billowing, to the end of the road to catch a bus to a friend's house. Her parents had been informed of this possible scenario and were willing to let me stay until I had got things sorted out. My plan was to move to London and find the life that I knew was out there waiting for me but would not be found on the streets of this small northern town.

As I was drinking hot sweet tea and re-telling the events that led to my outcast state the telephone began to ring. 'It's for you, it's your father,' my friend's mother called through to the kitchen where I was still in the process of thawing out from the reaction of my parents. If only I had kept my mouth shut. Nervously, I took hold of the receiver.

We have been talking and I'm coming to pick you up.

He began with

We should talk.

And then...

Are you a Queen?

I had no idea what he was talking about but could detect from the tone of his voice that whatever it was I had better not be one. As I sat at that formica topped kitchen table, the gathering place for every important family event I can ever recall, I tried hard to numb myself to the situation, to the disappointment etched on the faces of my parents. Martel (2003) notes:
Will I be understood when I say that sometimes numbness can hurt?

That you don't want to feel because what you feel will be pain, so
you try not to feel, and just sit there, immobile, numb, in pain?
(Martel, 2003, p. 40)

My father stood by the door and waited for me to give him the answer he wanted

No!

How do you know that you are...?

He avoided saying it.

Have you had sex with a girl?

Yes.

I lied.

Have you had sex with...

Again, he couldn't bring himself to say it, to name me.

No

I lied.

This was the last time my sexuality was ever mentioned. It was tiptoed around like I had some terminal illness, that if named would rear up and consume us all with one fell blow. The lesson I learnt from my initial revelation was one of how language and discourse 'is always productive: It brings a situation into play, enunciates evaluations of the situations and extends action into the future' (Denzin, 1997, p. 37). Looking back, what went and remained unsaid is far more descriptive and meaningful to my interpretation of that situation. I knew through their silence that my parents were holding out for a reversal, for a change of mind. In giving voice to what lay on the inside I had created myself as I wanted to be seen and heard. In their silence they were unseating that creation and hoping for an erasure of its possible existence

At 17, the spectre of AIDS stalked and prowled and I was told to have unprotected straight sex is risky but to have gay sex=death and I sought refuge in boy–on boy fiction to stave off desire and keep a distance from what had been constructed by others, for me, as a dangerous twilight world.

At 18, just when I thought it couldn't get any worse, it did...along came Clause 28 of the Local Government Act, that sought to stop local authorities from intentionally promoting homosexuality or promoting the teaching of the acceptability of homosexuality as a pretended family relationship.

At 19, I avidly read the literature that historically chronicled Queer experience longing to make contact.

At 20, I consumed autobiographies of famous Queers, thrilling in their daring resistances to orthodox heterosexuality and found between those pages what my parents had feared all along.

At 21, I waited and waited and waited for sanctuary from the coerced smile, the sideward momentary look, the furtive gaze.

Seeking refuge 2

At 22, seeking refuge I fled to Japan and in the backstreets of Tokyo I traded off the novelty of being a long way from home. Trawling my whiteness around the bars and backrooms, I acted out being the Other's Other. I felt REAL for the first time in my life. In the arms and between the legs of tattooed, leather capped Japanese lovers was how I came to perform resistances to the everyday architecture of heterosexuality. Seeking out and hunting down recognition in darkened bars I sought refuge from the autobiographical framing of my youth, I became momentarily detached from the emphatically masculine narratives of my parents worldh. I, reinscribed and enacted, fetishized the ritualistic scripts of the privileged and the represented. It was from this fiction of self that I endeavoured to queer the innumerable forms of social discipline that had been inscribed in, on and by my body. Surrendering to the strangeness of being simultaneously subject and object, I came to a better sense of understanding of the ways of belonging. It was in Dionysian moments of passion; in the fleeting madness of desire I came to experience a reforming power. Within the moment of my little deaths, as the world fell away and my mind was silenced, I came to know the violent truth of realizing a new landscape of promethean potentialities ... in the saunas and bathhouses, the disruption, immediacy and excess of ... trading off, passing by, giving up putting out, taking on, watching and of being watched, of playing with and of being played, of hands and mouths and arms, of holding out and taking in, of backs rolling and thighs flipping, of stretching and caressing, of lips and tongues and teeth, of kissing, licking, nibbling and biting, of mouths urging me on to being more than I had ever imagined was possible or could be pleasurable and coming to feelings of indeterminacy, fragmentation, discomfort and a sense of excitement. I felt increasingly more at home than ever before.

From my pleasures I re-remember the sense of hope that came from a feeling of being disconnected and for the first of recognising how the ways of my belonging were/ are intimately connected to the feeling and acting out the possibilities of the ways of my body. I am reminded how I am supposed to live alone, fall in love, to consume and be consumed. I have turned to seek refuge in the repetitions of my being: male, camp, butch, white, a rice queen, a dinge queen, a faggot, a queer, a sissy boy, a bottom seeking his top. I have throughout time and in places past hunted out and reclaimed places for the ways of fugitive bodies. I will continue to remake from the normative discourses of practice that sanction how affect, pleasure, power and identities are experienced. I have in the ways of my body become better equipped to resist being disassociated, dis-identified and differentiated by the normative and the stigma of male feminization is an enduring condition that has actively discredited how and who I claim to be. I am routinely located in hegemonic understandings of gender and sexuality that focuses on my enacted passive role taken within imagined sexual acts. My body has encountered the strange landscapes of the pleasures and prohibitions of what queer bodies are and what they can do; I have come to my practices of self. Looking back, I us, have travelled beyond being in my outcast state, I am becoming a refugee from the streets and my skin where I once felt community. I am not sure how or where they exist anymore. Where I can go to find a fuck, a body that I can trade off, pass by, give up put out, watch and play. I am no longer certain of hands and mouths and arms, of holding out and taking in, of backs rolling and thighs flipping, of stretching and caressing, of lips and tongues and teeth, of kissing, licking, nibbling and biting, of mouths urging me on to being more than I had ever imagined was possible or could be pleasurable. I am coming to feelings of indeterminacy, fragmentation and discomfort.

NOTES TO THE PLAY

[1] But it's not possible to continue in this way!
I do not understand anything. Why you write in this way!
But you do not put your foot in my shoes. Think about how for me is frustrating to continue to interrupt and not understand and continue to ask. And to continue to translate, also word by word, and understand nothing. It is frustrating!!! Perhaps it is better that we stop, that we give up writing together.
[2] No, this is not the 'Maschio Latino.' This is more like the 'Maschia isterica.'
[3] A person's name or surname followed by 'culo' (ass) is an Italian insult used mostly between young men; it is a way of advertising someone's homosexuality, or of outing them against their will.

REFERENCES

Adams, T.E. & Holman Jones, S. (2011). Telling stories: Reflexivity, queer theory, and autoethnography. *Cultural Studies – Critical Methodologies, 11*(2), 108-116.

Armstrong, D. (2003). *Experiences of special education: Re-evaluating policy and practice through life stories*. London: Routledge Falmer.

Dubois, W. (1973). *The souls of black folk*. Millwood, NY: Krauss-Thomson Organisation.

Duncan, D. (2005). *Reading and writing Italian homosexuality: A case of possible difference*. Aldershot & Burlington: Ashgate.

Ellis, C. (1997). Evocative autoethnography: Writing emotionally about our lives. In W.G. Tierney, & Y.S. Lincoln (Eds.), *Representation and the text: Re-framing the narrative voice* (pp. 115-142). Albany: State University of New York Press.

Ellis, C. (1999). Heartful autoethnography. *Qualitative Health Research, 9*(5), 669-683.

Grace, A., & Hill, R.J. (2001). Using queer knowledge to build inclusionary pedagogy in adult education. In R. Smith, J.M. Dirkx, P.L. Eddy, P.L. Farrell, & M. Polizin (Eds.), *Proceedings of the*

42nd Annual Adult Education Research Conference (pp. 145-150). East Lansing: Michigan State University.
Johnson, R. (1996). Sexual dissonances: Or the 'impossibility' of sexuality education. *Curriculum Studies, 4*(2), 163-189.
Mair, D. (2010). Fractured narratives, fractured identities: Cross-cultural challenges to essentialist concepts of gender and sexuality. *Psychology and Sexuality, 1*(2), 156-169.
McLaren, P. (1999). *Schooling as a ritual performance: toward a political economy of educational symbols and gestures*. Lanham: Rowma & Littlefield.
Marineau, Rene F. (1989). *Jacob Levy Moreno 1889-1974. Father of psychodrama, sociometry, and group psychotherapy*. London/New York: Routledge.
Moreno, J.L. (1964). *Psychodrama* (Vol 1). New York: Beacon House.
Plummer, K. (1992). *Modern homosexualities: Fragments of lesbian and gay experience*.London: Routledge.
Savin-Baden, M., & Fisher, A. (2002). Negotiating 'honesties' in the research process. *British Journal of Occupational Therapy, 65*(4), 191-193.
Schrag, C.O. (1986). *Communicative praxis and the space of subjectivity*. Bloomington: Indiana University Press.
Schratz, M., & Walker, R. (1995). *Research as social change*. London: Routledge.
Sedgwick, E.K. (1990). *Epistemology of the closet*. Berkeley: University of California Press.
Squires, N. (2010). Silvio Belusconi: 'My passion for women is better than being gay.' *The Telegraph*, 2 November 2010. Available at: http://www.telegraph.co.uk/news/worldnews/europe/italy/8105684/Silvio-Berlusconi-My-passion-for-women-is-better-than-being-gay.html.
Tierney, W., & Lincoln, T. (1997). *Representation and the text: Reframing the narrative voice*. Albany: State University of New York Press.
Watson, C. (2008). *Reflexive research and the (re)turn to the baroque: Or, how I learned to worrying and love the University*. Rotterdam: Sense Publishers.

Angelo Bennozo
University of Valle d'Aosta
Italy

Neil Carey
Manchester Metropolitan Universoty
Manchester, UK

Tarquam McKenna
Faculty Arts Education and Human Development
Victoria University
Melbourne, Australia

Mark Vicars
Faculty Arts Education and Human Development
Victoria University
Melbourne, Australia

GENEVIEVE NOONE

RHIZOMATICS AND THE ARTS

Challenging conceptions of rural teaching

INTRODUCTION

> If we cannot imagine things differently we will not be able to bring about any alteration in our circumstances. (O'Neill, 1995, p. 152)

For the past 150 years metrocentric research into rural education in Australia has produced "evidence" of the deficits in rural education. Many of these studies have, unknowingly, silenced rural voices by creating questions and representing data in the hegemonic, metro-centric language of urban based research. The study discussed in this paper (Noone, 2007) was an attempt to bring rural voices to the centre and present an alternative discussion of the "problem" of rural education: one which acknowledged the nature of rural place and the importance of teacher-place relations in rural education. The research explored the nature of the relations between place and becoming-teacher for five graduate teachers in rural schools in northern New South Wales, Australia. I was acutely aware of the plethora of studies into rural education which described the geographic isolation and culture shock experienced by graduate teachers. But where were their *stories* of place? Where were their *representations* of their experiences? What was their *lived* experience of place? And how could I "access" these experiences such that the data collected would enable me to portray them in ways that could lead to new perspectives on, and insights into, this phenomenon?

To avoid the reproduction of deficit language the study supported the participants (rural teachers) in using the arts to express their experiences of rural teaching. This resulted in the creation of artefacts with depth and breadth: some created by the participants; others created by the researcher as part of the analysis. In the original report the usual data tables and sample extracts were replaced by image/text representations; images overlaid with transparent sheets of text; pockets of multiple text/image artefacts; and "pull-out" pages and booklets of the participants' words. These data and researcher interpretations were then analysed using Deleuzo-Guattarian rhizomatic notions of becoming and place (Deleuze & Guattari, 1988): notions which challenge the linear, arboreal nature of much western research, and, instead, promote complex analyses and multiple understandings social phenomena. These conceptual notions challenge the deficit model of rural teaching and ask the audience to reconsider rural places as sites for generative models and understandings of rural teaching.

M. Vicars et al. (eds.), Discourse, Power, and Resistance Down Under, 77–94.

RURAL TEACHING

The growing body of research into rural education in western societies over the past half a century (Boylan & McSwan, 1998; Sharplin, 2002) tell us that it is difficult to attract teachers to, and retain teachers in, many rural areas (Committee for the Review of Teaching and Teacher Education, 2003; N. Hatton, Watson, Squires, & Soliman, 1991; Lonsdale & Ingvarson, 2003; Ralph, 2002; Sharplin, 2002; A. J. Watson & Hatton, 1995; Yarrow, Ballantyne, Hansford, Herschell, & Millwater, 1999). It tells us that students in rural areas leave school earlier and so do not have the same level of education as students in other places, and that for students who do complete the senior levels of schooling, often their outcomes are lower than their metropolitan counterparts (Hillman, Marks, & McKenzie, 2002; Lyons, Cooksey, Panizzon, Parnell, & Pegg, 2006; Marks, Fleming, Long, & McMillan, 2000). Some researchers suggest a link between the high teacher turnover and the lower retention and achievement levels of students (Hatton et al., 1991; McConaghy, 2002; Yarrow, Herschell, & Millwater, 1999). All this indicates that there are issues in rural education that need to be addressed in terms of both equity for rural students and the experiences of teachers in rural areas.

One of the tasks of the study, through the representations of the lived experiences of the teachers, was to allow particular rural places to 'speak back.' In Griffith's (2002) analysis of the language used to describe the rural in Australia, not only is the inland *other* than what is me and my experience, it is also 'inconvenient, uncomfortable, insecure, unproximate'. However, it is not just everyday language that moves the rural to the periphery. In reflecting on the rural in academic discourse Daniela Stehlik (2001, p. 32) notes that 'the academic language of cultural criticism has tended to "solidify dichotomies" between the urban and the rural, treating the urban as the central space from which language … emanates, and the rural as an abstract 'other'. This criticism of the language used in research on rurality as being metrocentric reflects an earlier, broader critique by David Geoffrey Smith (1997, p. 3) who noted that:

> in the West, most social theory since the turn of the century has been generated in urban, highly industrialised environments … [and] such a site for the generation of theory and policy must inevitably ensure social outcomes that are driven by the ambiances, rhythms, and tone of their situational origin.

So it is not simply that the language of rural research has treated the rural as 'other,' but, according to Smith, that research generated by researchers living and writing in urban environments necessarily reflects the rhythms of urban life. Both Stehlik and Smith are suggesting that if we want social theory that represents and is appropriately formulated for rurality we must address the site of the research, and in particular the language which the research is expressed in. The study discussed in this paper attempted to move the rural from the periphery to the centre by focussing on the lived experiences of teachers in rural schools, and using their words as much as possible in the representations of their experiences. The use of

Deleuzo-Guattarian notions in analysing the data avoided the use of hegemonic themes often applied to rural teaching, replacing them with abstract concepts aimed at challenging current understandings of rurality and teaching.

A DELEUZO-GUATTARIAN RHIZOMATICS OF PLACE AND BECOMING

Deleuze and Guattari (1988) choose the rhizome, an underground, tuberous growth, to illustrate their notion of *multiplicity*: of no object and no subject, only events. A rhizome, unlike a root, runs horizontally beneath the surface of the earth. Unlike roots, which travel linearly, a rhizome is neither linear, nor circular, but multiple; sending out roots and shoots from its numerous nodes. It is this process of "sending out" multiple lines of connection from multiple nodes that is the core of the notion of multiplicity.

Rhizomes are characterised by connection and heterogeneity: 'any point of a rhizome can be connected to anything other, and must be'. As a multiplicity, a rhizome has no points or positions, only 'determinations, magnitudes, and dimensions,' and when it is broken 'it will start up again on one of its old lines, or on new lines'. A rhizome is made up of lines, rather than points: 'lines of segmentarity' (rupture) as well as 'lines of deterritorialisation down which it constantly flees' (Deleuze & Guattari, 1988, pp. 6-7); lines which pass between things, between points. It is this movement – connecting, segmenting, fleeing – that is central to the notion of multiplicity. Rather than an either/or, the focus is on the *movement*. It is movement which creates connections. As newly appointed teachers in rural places the teachers in the study were performing new roles in unfamiliar places; of necessity they made connections with both the familiar and unfamiliar; they experienced ruptures, and movements in who they were – in their relations to and in place. The concept of the *rhizome,* of movement and flows, of connections and disconnections, provided me with a perspective from which to allow the teachers' descriptions of lived experience to be engaged with without reduction to (predefined or emergent) categories. The notion of *rhizomatic becoming* provided a means of conceptualising the unpredictable, variable and messy process of learning to teach, of *becoming-teacher*.

Over the past decade many educational researchers have begun to use the philosophical works of Gilles Deleuze and, what is often referred to as, Deleuzo-Guattarian rhizomatics, as a way to rethink what we know about education and, armed with these new knowledges, to address educational issues from new perspectives. The body of research encompasses such diverse areas as creativity and the arts (Lines, 2005), science and technology education (Gough, 2005), environmental education (Hardy, 2006), pedagogies (Semetsky, 2006), educational policy (Honan, 2004), research methodologies (St. Pierre, 1997a; Youngblood Jackson, 2003; Honan & Sellers, 2006; Honan, 2007), and educational philosophy and the sociology of education (Semetsky, 2004; St. Pierre, 2004; Hickey-Moody, 2005; Zembylas, 2007). In particular Deleuzo-Guattarian rhizomatics have been used to rethink both concepts of the environment (place) in the curriculum, and bodies (teachers and pupils) in classrooms. Noel Gough (2005, p. 639), writing in regard to science education, uses the notion of rhizomes to make connections between science, literature and the arts, arguing that nature is a cultural experience, rather than something that can simply be represented structurally by arborescent scientific concepts. He argues that creating rhizomes between 'texts of science education, media reports, social studies and histories of science, and SF [science fiction]' can enable different, generative ways for science educators to engage with the social context of nature. Michael Zembylas (2007) also chooses to utilise rhizomatics to enable a different way of engaging with educational issues, in this case, with the issue of bodies in the classroom. Asserting that pupils and teachers bodies have been hidden, due to both the mind-body split in western thought, and perceptions of the body as a problem (the sexual as a sin), he proposes that, rather than perceiving of corporeal bodies, we address the abstract notion of the body as assemblages of movements with certain capacities. With this perception Zembylas suggests that we can begin to look at how 'the body' is engaged in learning in terms of energies and affects. Similarly this study built upon both the rhizomatic notion of place as social, and the rhizomatic notion of body as an assemblage of movements; with an emphasis on the affects of deterritorialisation and reterritorialisation.

DETERRITORIALISATION

> deterritorialisation must be thought of as a perfectly positive power that has degrees and thresholds. (Deleuze & Guattari, 1988, p. 54)

Deterritorialisation is 'the movement by which "one" leaves the territory'; where the territory, rather than being a milieu, is an *act* of territorialising milieus and rhythms (Deleuze & Guattari, 1988, p. 508). Everything has its own territory: its own acts which territorialise its surrounds. The study sought to *deterritorialise* rural teaching research by building on research which uses theories and philosophies of place, and challenging hegemonic perceptions of rural teaching which project rural teaching in deficit terms. Often the challenges of unfamiliar places, the distance from urban centres, and frequent movement between schools, are seen as negative, rather than deterritorialisations, which can have a positive power. The 'flipside or complement' of deterritorialisation is reterritorialisation. There is never deterritorialisation without reterritorialisation, but reterritorialisation is not a returning to the same territory as prior to deterritorialisation. Once deterritorialised "one" is never the same as prior to deterritorialisation. The territory has changed.

ARTS AND THE LIVED EXPERIENCE – CREATING DIFFERENT VOICES

> if the purpose of research is the creation of new knowledge, then the outcome is not merely to help explain things in causal or relational terms, but to fully understand them in a way that helps us act on that knowledge … arts researchers at all levels of education … have to be confident that by

> following different, yet complementary pathways, we can create important new knowledge. (Sullivan, 2006, p. 26)

The creative arts are able to evoke and capture sensory aspects of experience in ways that are not possible in talk or writing. After visiting an installation in an art gallery which moved her greatly, educational researcher Ardra Cole (2004, p. 16) wrote:

> If I were to make a difference through my work I could no longer rely on the very limited power of flaccid words and numbers. I needed words plump and dripping with life juice, compelling and evocative images, representations that drew readers and viewers in to experience the research "text" ... It was my job to more fully portray the complexities of the human condition to broader audiences, and to invite even an approximation of the kind of holistic, full-bodied engagement I had experienced through art. Research, like art, could be accessible, evocative, embodied, empathic, provocative.

In regard to explorations of teacher development, Patrick Diamond and Carol Mullen (1999, p. 20) argue for the use of arts-based methods of inquiry as they 'make experience more accessible, concrete, imaginable, and affecting.' Elliot Eisner (1993, p. 6) suggests that:

> since forms of representation differ, the kinds of experiences they make possible also differ. Different kinds of experiences lead to different meanings, which, in turn, make different forms of understanding possible.

The challenge for the study was deciding which forms of the creative arts would best facilitate both the teachers' expressions of their lived experience, and the researcher's analysis and representation of the same.

THE GIFTING OF THE DATA

To explore the five graduate teachers' experiences of both teaching and rural place I planned to visit them at their school once during each of the four school terms during their first year of teaching. During each visit I would observe the teacher in the classroom for a short period, then interview them as soon as possible following the observation. Following Leslie Rebecca Bloom's (1998, p. 19) suggestion for exploring lived experience I decided to simply ask each participant to "tell me about ...." The study asked *what is the nature of the relationship between place and teaching?* And the participating teachers were asked: *tell me about this place; tell me about teaching here;* and *tell me about yourself as teacher.*

In my goal to rethink *teacher* and *rural* as *becoming* and *place*, and in doing so challenge the deficit, metro-centric views of rural teaching by providing creative ways of representing the experiences of rural teachers, I also chose to invite the participants to a two-day creative arts workshop which was held in the break between the first and second school terms. The workshop engaged the participants in object collage, sketching, mixed media representations, soundscapes and dramatic performance. With each activity the participants were free to choose to share as little or as much as they chose of their creative work. Max van Manen (1997, p. 54) reminds us that 'originally, "datum" means something "given" or "granted."' Any artefacts the teachers shared with me were gifts which they chose to give.

Just as Nadia Wheatley and Ken Searle (2003) begin their respective creative processes with objects from place, and in the same way as Hannah Hinchman (1997) pays attention to the detail of the things in her place, I asked the teachers to bring objects, both natural and manmade, from their places to the workshop for the purpose of creating an object collage of teaching in a rural place. Following Margaret Somerville and Patsy Cohen (1990), who tell the story of the Aboriginal woman who keeps in a box the things that are important to her, I gave each participant a box to keep their objects in, along with other objects they might collect over the course of the year.

In the discussion of the data and interpretations in the following sections I will refer to some of the interview data, and only one of the art events, the object collage. Space does not allow for discussion of the other artefacts. The object collage has been chosen as the process of creating the collage elicited detailed and indepth descriptions of the lived experiences of both place and becoming-teacher.

EXPLORING PLACE AND BECOMING

In the process of working with and through the artefacts gifted by the participants (both the artefacts, and spoken words), I created text-image artefacts, many linking and/or juxtaposing images and text. This creative process opened up the liminal space I needed to begin to conceptualise "rurality" and "teacher" as movements: as speeds and slownesses; as ever-changing multiplicities.

Place relations

Each of the rural places in the study was quite different from the others. They were unique. And each of the teachers noticed different things about their place. These interview extracts tell of the teachers' responses to their unfamiliar places:

it's flat

it's cold

it's welcoming

it's far enough away to have your own time and your own events

it's a busy place

I did the rim of my tyre in

David Abram (1996, p. 50) insists that our responses to place are dependent upon our ability to be receptive to place, and our ability to be creative in responding. Some responses reveal comparisons to previous experiences:

my home town's an older community

I went through a small school when I was young

it reminds me of my home town when I was young

Other responses compare the lived experience to prior expectations:

it's not that far away

there are a lot of people from out of town ...

it's just amazing for a town this size

I didn't see this town as that sort of place

the quiet life

except for last week

... murder ... robbery ... house fires ... deaths

this town is not as bad as people make out

And someone noticed that not everyone is the same:

I have noticed the Aboriginal English...

I notice ... the property owners ... how good their English is

you've got the two divides

we have several different church groups

community groups ...

some are involved with horses ...

the other side is into the sport and tennis

you've got the higher class

and you've got your lower class

The stories told by the teachers were stories of adjustment, of movement, of relations with place. Most importantly, they were stories of the teachers' bodies in unfamiliar (yet in some ways familiar) places.

While the teachers' descriptions of their places told of encounters with both the human and nonhuman in place, and of their receptivity and adjustment to the unfamiliar (yet at times familiar), the objects were able to elicit descriptions of more intimate relations with and in place. At the workshop I asked the teachers to talk about the objects they had brought along *before* they began making the collage. One of the teachers brought along a bindii. See Figure 1 for the story told about this bindii.

In this story the teacher describes what could be a seemingly unimportant part of the experience of place – a weed he has not come across before. This lived-experience description tells us much about this teachers' life in a rural town. His description intermingles the past and the present, his family, his amazement, his pain. It is this attention to the obvious that enables us to gain a better understanding of ourselves (our becomings) in the world (in place).

In presenting their objects the teachers did not simply describe their objects or talk about the physical places they were from. Rather, each of the teachers, without any direction from myself, talked about their relation to the objects, introducing members of their family and friends, and discussing their teaching and teaching relationships. In speaking about their objects the teachers shared stories of their intimate selves: stories of wonder, anger, joy, frustration, fun, reminiscences, being overwhelmed, puzzlement, and finding sanity. The objects are representations of deterritorialisations, and reterritorialisations; of becoming-teacher.

The teachers also linked their objects to the history of their places and their own personal histories. There was an awareness of the past as present in and a part of the here and now; an awareness of multiplicities and assemblages from the past creating lines of flight in the present, lines of flight that are a part of their becoming-teacher. For the teacher whose room overlooked the town's famous football field, and who would walk his class through the historical cemetery, these places and their histories were a part of his experience of place in the present, and they connected his becoming-teacher with the histories of his pupils' families and community. A teacher's planner was a connection to other family members and their becomings, and the torn off calendar pages were a connection to very recent challenges of becoming-teacher. These connections were part of, and created rhizomes of, becoming-teacher. The past, present and future are all contained with the rhizome; they are all a part of becoming.

Place is an experience as much as it is a location or site. The teachers' descriptions of their places and their objects consisted of interminglings of things physical and non-physical, human and non-human, animate and inanimate. Deleuze and Guattari (1988, p. 21), in explaining the nature of rhizomes, write that 'what is at question in the rhizome is a relation to sexuality ... to animal, the vegetal, the world, politics, the book, things natural and artificial ... all manner of "becomings." In arguing for a role for place in the curriculum David Geoffrey Smith (1997, p. 4) contends that:

> Whatever the pedagogy of place may be, it has little to do with a warm cosy relationship with an imagined nature, and perhaps more to do with the courage to befriend one's own mortality in the midst of the ongoing project of self understanding.

now I don't know if you know what these are

and I'd never experienced them until I
got down to where I am
I don't remember them back where I grew up
and they didn't have them here

a couple of times we were walking out barefoot
and
trod on some of these
things and
lived to remember

they're painful

they stick in stick in your heel

and thongs aren't safe

I've actually got some still in my shoes
my youngest daughter she often goes outside
with just a nappy on and she scrapes her
bottom along the ground
and then she comes back inside and then just
rubs her bottom on the carpet
so we'll be coming down stairs and *ouch!*
heel straight onto it
when that happened a couple of weeks ago
it was a double pronged one
so I had two prongs go straight up in my heel
that's something I won't forget about my place

Figure 1. Story told by participant about the bindii.(Note: The term "thongs" refers to flat, backless, rubber footwear commonly worn by Australians)

A pedagogy of place requires a self understanding – an awareness – of one's relations with place, and this requires courage. In creating an awareness of their places the teachers were indeed becoming aware of themselves in relation to place. And in doing so they were becoming aware of their becoming-teacher.

TEACHER BECOMINGS

> the experience [of rural teaching] is more than learning …
>
> important aspects of good rural teaching and good rural leadership belong in the realm of the experience of place and cannot be contained or prescribed within the teacher learning curriculum. (McConaghy, 2006, p. 51)

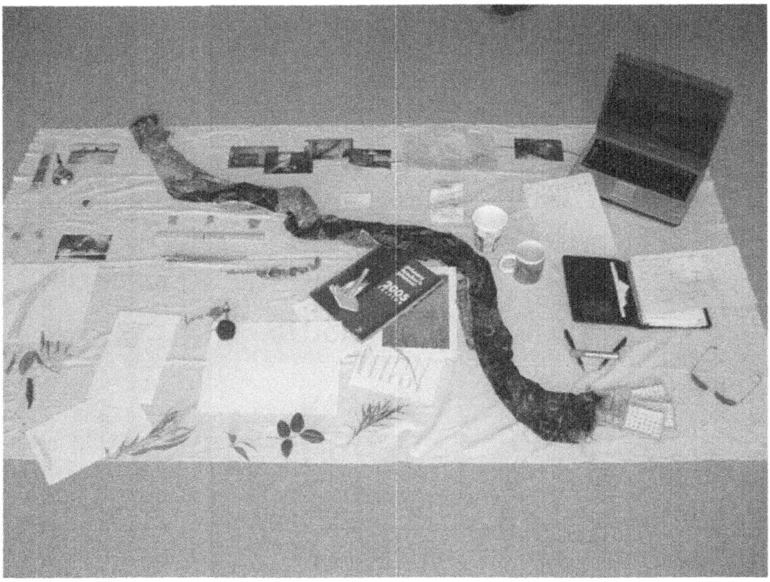

Figure 2. Photograph of completed object collage

After creating the collage, depicted in Figure 2, the participants were invited to share their thoughts about it. The ensuing discussion told of both deterritorialisations and re-territorialisations. The teachers spoke of movement (flow, drawing close) and blockages (weight, loneliness); of lines (diagonals) and points (cross-section); of trajectories (coming back, pointing towards) and impasses (strings attached, isolate). They spoke of rupture (shock) and rejoining (healing). They spoke of the difficulty of moving between the environment and teaching: place and teaching being divided by the cross-secting diagonal of fabric.

The collage depicted the complexity of teaching – so many different elements of place and teaching to move with and in-between.

Each term, on my visits to the teachers in their schools, I asked them to *tell me about teaching in this place*. The teachers' tales tell of challenges and successes, and joys and frustrations. There are no prescriptions for graduate teachers to follow in their becomings-teacher. Each individual is a unique assemblage of multiplicities, connecting with other unique assemblages (human and non-human, animate and inanimate) of and in their places of teaching. Their stories tell how they created their own individual lines of flight to manage the impasses they met:

how they chose to –

compromise,

get organised,

make a deal,

give up, and

not always deliver 100%;

and how they got –

more relaxed,

resigned,

greater memory,

and comfortable.

Their lived-experience descriptions describe a part of their becoming-teacher. Deleuze and Guattari (1988) describe becoming cartographically, proposing that there are two axes of becoming – longitude and latitude; longitude being *relations of movements*, and latitude being *capacities to affect* (where *affect*, rather than being a personal feeling, is the capacity to affect other assemblages and to be affected by them). Becomings exist on this plane designated by longitude and latitude, and a body on this plane is defined by:

> material elements belonging to it under given *relations* of movement and rest, speed and slowness (longitude); [and] the intensive *affects it is capable* of at a given power or degree of potential (latitude). (Deleuze & Guattari, 1988, p. 260, emphasis added)

The teachers' descriptions of teaching tell of their relations and capacities; of the speeds and slownesses of the movements of their becomings, and in relation with pupils, colleagues, syllabuses, practices, policies and rules and the relations between and among them; and of their capacities to affect and be affected by these assemblages. And while we can distinguish between relations (movements) and capacities (affects) they are intimately entwined. The teachers described ruptures that required adjustments on their part; blockages and impasses that hindered their teaching; and lines of flight they took to enable their becomings-teachers to be generative, rather than the alternative which would have been to allow their lines of flight to be obstructed or become lines of destruction (Deleuze & Guattari, 1988, p. 210), which may have led to a decision to leave the teaching profession.

It is common today in education circles to hear that teachers "survive" (or don't "survive") their first year of teaching. However, survival is not necessarily a positive movement. Survival in the context of the study would mean merely that the teachers returned to their schools the following year to continue teaching. And they all did. But who is to say that the lines of flight the teachers took to *survive* will not ultimately be destructive, or alternatively productive?

In their first year of teaching the teachers adapted things to suit themselves, and adapted themselves to suit things. Their capacity to adapt to the places in which they found themselves is, according to David Abram (1996) directly related to their capacity to be receptive to place. Becomings are always in some place. The teachers created spaces in these places for their becomings-teacher, "smooth spaces" (as opposed to striated space, see Deleuze & Guattari, 1988, p. 380) which enabled both the relations of movements, and the capacities to affect and be affected, that were necessary for a becoming-teacher.

Becoming-teacher is a risky business. It involves letting go of old, comfortable understandings of who we are, and taking on new identities (Britzman, 2003; Danielewicz, 2001). And in unfamiliar places, there can be even greater destratification. The high rate of teachers who leave teaching within the first five years of graduating (Committee for the Review of Teaching and Teacher Education, 2003, p. 87) suggests that many becomings-teacher do end in chaos or a void.

RURAL TEACHING AS PLACE AND BECOMINGS: AFFECTS, RELATIONS & BODY

> … we avoided defining a body by its organs and functions, we will avoid defining it by Species or Genus characteristics; instead we will seek to count its affects. (Deleuze & Guattari, 1988, p. 257)

In the representations of teaching we can see the struggles of the teachers to create the places and spaces they desired for their becoming-teacher. The smooth spaces of their becomings – the spaces of movement and relations – were full of both challenges and successes. In the collage the striated spaces of programming and recording, of the planner, the budget, yard duty, and pens, were set alongside the movements of the river and bursting seeds and aging bark, the long walk and the short walk; and were mediated by the flow from books to music to arts. Their relationships with their pupils – the sounds of their conversations and their attempts to create order from chaos – were paramount in their concerns. The teachers took lines of flight that deterritorialised and reterritorialised them. The capacities of teacher assemblages' were diminished and augmented as they affected and were affected by other assemblages – by their pupils and colleagues; by school structures, rules, programs, policies, and the geography of place; and by their perceptions and relationships with time.

A becoming-teacher is an act of approximating the relations of movements and capacities of a teacher, that is, developing the affects of a teacher. In notes on his translation of Deleuze and Guattari's work, Brian Massumi (1988, p. xvi) writes:

> L'affect
> is an ability to affect and be affected.
> It is a prepersonal intensity
> corresponding to the passage from one experiential state of the body to another
> and implying an augmentation
> or diminution
> in that body's capacity to act.

At various times during the year the teachers' capacities to teach (their affects) were augmented and diminished by the speeds and slownesses of the movements of their becomings, and by the affects of other assemblages upon them. The teacher who was treated as an aide by a colleague, the teacher whose classes kept disappearing for sport and play rehearsals, the teacher who didn't know of requirements regarding his students' external exams, the teacher who had to take some of the principal's students in her classes, and the teacher who couldn't find the time to get his bus licence – in all these instances the capacities of these teachers to act (to teach) were diminished. However, there were many instances where they were augmented – the teacher who was assisted to settle in by a colleague, the teacher who was able to take his pupils by bus to "do IT stuff," the teacher who decided that it was not necessary to "always deliver 100%," the teacher who had so many other people for support . Deleuze and Guattari (1988, p. 257) write:

> We know nothing about a body until we know what it can do, in other words, what its affects are, how they can or cannot enter into composition with other affects, with the affects of another body, either to destroy that body or to be destroyed by it, either to exchange actions and passions with it or to join with it in composing a more powerful body.

In analysing the teachers' lived-experience descriptions I was exploring what the affects of the teachers were: how they were affected and how they affected others. On each visit the teachers were able to tell me about both successes and challenges; frustrations and joys. At any one time it was difficult to say whether any of the teachers were taking lines of flight that would end in destruction; or if they were indeed soaring. At each interview they spoke of both descents and ascents.

CONCLUSION

> *One answer! One explanation! One cause! That's what we yearn for …*
>
> But what if the human condition isn't like that? What if some of the questions we face actually are complex? What if most events that occur in human systems are the result of many factors? What if mystery and ambiguity are things we simply have to live with? I'm inclined to agree with the Indian mystic, Krishnamurti: 'Freedom from the desire for an answer is essential to the understanding of a problem.' (Mackay, 2005)

Research into rural education over the past half a century has suggested that rural education is *a problem*: that it is difficult to attract and retain teachers and that student achievement and retention are lower than in the cities. For the most part the research has constructed the rural teacher as a temporary interloper into the life of rural schools and their communities. Teachers are presented as individuated subjects and the rural schools and communities as objects and subjects of the experience of teaching. Much of the research presents rural schools and communities as geographically isolated and "home" for the teachers as somewhere else. The difficulty in recruiting teachers to rural schools has resulted in the situation where many graduate teachers are appointed to these schools: it is where many teachers obtain their first experiences of teaching their own classes; and it is where important events of becoming-teacher occur.

I have suggested in this paper that part of *the problem* of rural teaching is the deficit conceptions presented in much of the literature and I have proposed the appropriation of Deleuzo-Guattarian rhizomatics as a way to rethink rural teaching in more constructive and creative terms. From the perspective granted by rhizomatics becoming-teacher is a process and teacher is an event, as opposed to teacher and teaching as known and knowable subjects. Place, also, is a process, consisting of mutual, dynamic relations between the human and nonhuman, and the animate and inanimate. These conceptions not only disrupt conceptions of the rural teacher and rural teaching simply as problems of geographic isolation, and cultural shock. No longer is the teacher simply an object or subject, but a part of the processes of place in which s/he is teaching; affecting and being affected by place. Elizabeth St Pierre (2004, pp. 284–5) suggests that the usefulness of the ideas of Deleuze and Guattari are in the possibilities they enable for different ways of thinking and feeling and for creating lines of flight that enable different ways of being in the world; so, in the end, this is what needs to be asked of the study: what new thoughts and emotions does it make possible? and what different ways of being in the world does it enable?

This analysis enables a different understanding of *the problem* of rural education. For many teachers taking up teaching positions in rural areas in Australia, an awareness of their mutual relations with place is something that may need to be nurtured. Certainly the notion of a mutual, intimate relation with place is not addressed in much of the literature which makes, or reports on, recommendations regarding teacher preparation for rural teaching (Boylan, 2002,

2004; Human Rights and Equal Opportunity Commission, 2000; Sharplin, 2002; Vinson, 2002; Yarrow et al., 1999). However, many of these writings do emphasise the importance of preservice teachers spending time in rural places (that is, taking sensing bodies to dwell in place), as well as the importance of developing social relations; both of which have been suggested by the study as important elements in developing a sense of place, and as important in teacher becomings.

In reporting the findings of the first stage of the *Bush Tracks* project McConaghy, Lloyd, Hardy and Jenkins (2006, p. 27) write:

> both the contextual and affective dynamics of rural teaching were prominent in our teacher interviews … they overwhelmed teachers' responses in relation to the cognitive dimensions of good teaching … Pedagogical content knowledge needs to be able to accommodate the affective and the contextual dynamics of teaching.

This suggests that rural teachers themselves acknowledge that the intimate experience of place (the dynamics of the context and emotions of teaching) are equally as important as knowledge of how and what to teach. Other recent rural education research by Jo-Anne Reid et al. (2010, p. 269) has begun to address the role of "place" in rural teaching suggesting that:

> it is the practice of place that provides and produces social space, and the way in which these factors interact and interrelate that suggests ways in which rural social space can be rethought and represented in ways that do not produce symbolic deficit and cultural cringe.

Elsewhere McConaghy (McConaghy, 2006; McConaghy & Bloomfield, 2004) explores the intimacies of place through psychoanalytic analyses, and suggests that we need new perspectives which query commonly held beliefs about the negative effects of teacher mobility, and rural teaching in general, and that rural teaching may indeed be a place which is generative of valuable new knowledges and new possibilities. The study discussed in this paper continued this focus on the personal experience of place as part of the experience of rural teaching. Becoming-teacher always occurs in some place. We are always in place. And we are always in mutual relations with place.

Conceiving of rural teaching as movements and flows, as place and becomings, as relations within and between all manner of things, enables the focus to be shifted, from the individuated teacher, to places of teaching; places which consist of a myriad of relations which need to be understood and nurtured if teachers and rural communities are to be supported in creating equitable outcomes for rural education.

REFERENCES

Abram, D. (1996). *The spell of the sensuous*. New York: Vintage Books.

Bachelard, G. (1994). *The poetics of space*. Boston: Beacon Press.

Bloom, L.R. (1998). *Under the sign of hope: Feminist methodology and narrative interpretation*. Albany: State University of New York Press.

Boylan, C. (2002). *Reviewing rural teacher education: preparing and supporting quality teachers for rural New South Wales schools.* Discussion paper. Bathurst, Australia: Charles Sturt University.

Boylan, C. (2004). Putting rural into pre-service teacher education. Australian Association for Research in Education Annual Conference. Retrieved from www.aare.edu.au/04pap/boy04081.pdf.

Boylan, C., & McSwan, D. (1998). Long-staying rural teachers: Who are they? *Australian Journal of Education, 42*(1), 49-65.

Britzman, D. (2003). *Practice makes practice: A critical study of learning to teach.* Albany: State University of New York Press.

Cole, A.L. (2004). Provoked by art. In A.L. Cole, L. Neilsen, J.G. Knowles, & T.C. Luciani (Eds.), *Provoked by art: Theorizing arts-informed research*, Arts-informed Inquiry Series (Vol. 2, pp. 11-17). Halifax, Nova Scotia: Backalong Books & Centre for Arts-informed Research.

Committee for the Review of Teaching and Teacher Education. (2003). Australia's Teachers: Australian's Future – Advancing Innovation, Science, Technology and Mathematics – Main Report. Retrieved November 23, 2006, from www.dest.gov.au/sectors/school_education/ policy_initiatives_reviews/reviews/teaching_teacher_education/review_of_teaching_and_teacher_ education_home.htm.

Danielewicz, J. (2001). *Teaching selves: Identity, pedagogy, and teacher education.* New York: State University of New York Press.

Deleuze, G., & Guattari, F. (1988). *A thousand plateaus: Capitalism and schizophrenia.* (B. Massumi, Trans.). London: The Athlone Press.

Diamond, C.T.P., & Mullen, C.A. (1999). Art is a part of us: From romance to artful story. In *The postmodern educator: Arts-based inquiries and teacher development* (pp. 15-36). New York: Peter Lang.

Eisner, E.W. (1993). Forms of understanding and the future of educational research. *Educational Researcher, 22*(7), 5-11.

Gough, N. (2005). Shaking the tree, making a rhizome: Towards a nomadic geophilosophy of science education. *Educational Philosophy and Theory, 38*(5), 625-645.

Griffiths, T. (2002). The outside country. *Words for country: Landscape and language in Australia* (pp. 222-244). Sydney: University of New South Wales.

Hardy, J. (2006). Theorising "Postmodern Uncertainities" in education for sustainable development: Philosophical and methodological considerations. In N. Kyriakidou (Ed.), *International reflections of education and business* (pp. 49-63). Athens: Athens Institute for Education and Research.

Hatton, N., Watson, A., Squires, D., & Soliman, I. (1991). School staffing and the quality of education: Teacher stability and mobility. *Teaching and Teacher Education, 7*(3), 279-293.

Hickey-Moody, A. (2005). *Unimaginable bodies: Intellectual disabiity, performance and becomings.* Doctoral dissertation. University of South Australia, Adelaide, Australia.

Hillman, K., Marks, G., & McKenzie, P. (2002). *Rural and urban differences in Australian education.* LSAY Briefings Number 5. Melbourne, Australia: Australian Council for Educational Research.

Hinchman, H. (1997). *A trail through leaves: The journal as a path to place.* New York: W. W. Norton & Co.

Honan, E. (2004). (Im)plausibilities: A rhizo-textual analysis of policy texts and teachers' work. *Educational Philosophy and Theory, 36*(3), 267-281.

Honan, E. (2007). Writing a rhizome: an (im)plausible methodology. *International Journal of Qualitative Studies in Education, 20*(5), 531-546.

Honan, E., & Sellers, M. (2006). So how does it work? Rhizomatic methodologies. Presented at the Australian Association for Research in Education Annual Conference. Retrieved from www.aare.edu.au/06pap/hon06003.pdf

Human Rights and Equal Opportunity Commission. (2000, September 14). Rural and Remote Education Inquiry. Retrieved May 23, 2005, from www.hreoc.gov.au/human_rights/rural_education/

Lines, D. (2005). *Music education for the new millenium.* Oxford: Blackwell.

Lonsdale, M., & Ingvarson, L. (2003). *Initiatives to address teacher shortage.* ACER Policy Briefs – 5. Melbourne, Australia: Australian Council for Educational Research.

Lyons, T., Cooksey, R., Panizzon, D., Parnell, A., & Pegg, J. (2006). *Science, ICT and mathematics education in rural and regional Australia: State and territory case studies – A companion volume to the SiMERR National Survey*. Armidale, Australia: University of New England.

Mackay, H. (2005, March 20). Social disengagement: a breeding ground for fundamentalism. *Big ideas – Manning Clark Lecture*. Australia: Australian Broadcasting Commission. Retrieved from www.abc.net.au/rn/bigidea/stories/s1323906.htm.

Marks, G., Fleming, N., Long, M., & McMillan, J. (2000, January 23). Patterns of participation in year 12 and higher education in Australia: Trends and issues. Retrieved from www.acer.edu.au/research/LSAY/ExecSummary/execsum17.html.

Massumi, B. (1988). Notes on the translation and acknowledgments. *A thousand plateaus: Capitalism and schizophrenia* (pp. xvi-xix). London: Continuum.

McConaghy, C. (2002). *Situated pedagogies: Researching quality teaching and learning for rural New South Wales schools.* Discussion paper. Armidale, Australia: University of New England.

McConaghy, C. (2006). Transience and teaching: Place and the new psychoanalytic sociologies of teaching. *Education in Rural Australia, 16*(2), 47-57.

McConaghy, C., & Bloomfield, D. (2004). Beyond the line, beyond the self: flirting with rural teaching. *Journal of Curriculum Theorizing, 20*(3), 93-112.

McConaghy, C., Lloyd, L., Hardy, J., & Jenkins, K. (2006). Bush tracks: Journeys in the development of rural pedagogies. *Education in Rural Australia, 16*(2), 13-29.

Noone, G. (2007). *Place and becoming: A study of graduate teachers in rural schools.* Doctoral dissertation. University of New England, Armidale, NSW, Australia. Retrieved from https://e-publications.une.edu.au/vital/access/manager/Repository/une:1914?query=noone.

O'Neill, C. (1995). *Drama worlds: A framework for process drama*. Portsmouth, NH: Heinemann.

Ralph, E.G. (2002). Teaching in rural schools. *Rural Educator, 24*(2), 13-21.

Reid, J.-A., Green, B., Cooper, M., Hastings, W., Lock, G., & White, S. (2010). Regenerating rural social space? Teacher education for rural–regional sustainability. *Australian Journal of Education, 54*(3), 262-276.

Semetsky, I. (2004). Becoming-language/becoming-other: whence ethics? *Educational Philosophy and Theory, 36*(3), 313–325.

Semetsky, I. (2006). *Deleuze, education and becoming*. Rotterdam: Sense Publishers.

Sharplin, E. (2002). Rural retreat or outback hell: Expectations of rural and remote teaching. *Issues in Educational Research, 12*(1), 49-63.

Smith, D.G. (1997). The geography of theory and the pedagogy of place (editorial). *Journal of Curriculum Theorizing, 13*(3), 2-4.

Somerville, M., & Cohen, P. (1990). *Ingelba and the five black matriarchs*. Sydney: Allen & Unwin.

St. Pierre, E.A. (1997). Methodology in the fold and the irruption of transgressive data. *Qualitative Studies in Education, 10*(2), 175-189.

St. Pierre, E.A. (2004). Deleuzian concepts for education: The subject undone. *Educational Philosophy and Theory, 36*(3), 283-296.

Stehlik, D. (2001). "Out there": Spaces, places and border crossings. In *Rurality bites* (pp. 30-41). Annandale NSW: Pluto Press.

Sullivan, G. (2006). Research acts in art practice. *Studies in Art Education, 48*(1), 19-35.

van Manen, M. (1997). *Researching lived experience* (2nd ed.). Ontario: The Althouse Press.

Vinson, T. (2002). *Inquiry into the provision of education in New South Wales* (second report). NSW Teachers Federation & Federation of P&C Associations of NSW. Retrieved from www.pub-ed-inquiry.org/reports/final_reports/03/chapter7.html.

Watson, A. J., & Hatton, N. G. (1995). Teacher placement and school staffing. In L W. Anderson (Ed.), *International encyclopedia of teaching and teacher education* (pp. 606-611). Oxford: Pergamon.

Wheatley, N., & Searle, K. (2003, September 20). *Letting the landscape tell the story*. Presented at the ALEA Symposium, Parramatta, Australia.

Yarrow, A., Ballantyne, R., Hansford, B., Herschell, P., & Millwater, J. (1999). Teaching in rural and remote schools: A literature review. *Teaching and Teacher Education, 15*, 1-13.

Yarrow, A., Herschell, P., & Millwater, J. (1999). Listening to country voices: Preparing attracting and retaining teachers for rural and remote areas. *Education in Rural Australia, 9*(2), 1-12.
Youngblood Jackson, A. (2003). Rhizovocality. *Qualitative Studies in Education, 16*(5), 693–710.
Zembylas, M. (2007). The specters of bodies and affects in the classroom: A rhizo-ethological approach. *Pedagogy, Culture and Society, 15*(1), 19-35.

Genevieve Noone
School of Education
University of New England

SATOSHI SANADA

WRITING 'RACE' INTO ABSENCE? POST-RACE THEORY, GLOBAL CONSCIOUSNESS AND REFLEXIVITY

INTRODUCTION

This chapter discusses tenets of 'post-race theory,' a newly emerging theories that I am interrogating in my current PhD study on migrant teachers in Australia. Post-race theory remains underdeveloped and has been to be highly contentious. At this point, I treat post-race theory as a tentative and experimental project that is directed at questioning researchers' own preoccupations and presumptions about 'race.' While post-race theory seems to possess many tenets that require further consideration, I will limit my discussion to its critique of 'race' from the anti-foundationalist perspective, and how the development post-race theory can inform and is informed by the current works of global consciousness and reflexivity. My intention is to open a critical discussion and to demonstrate how post-race theory can potentially be an exciting and quite felicitous framework which can add to the existing bodies of work on 'race,' such as postcolonial theory and Critical Race Theory (CRT).

'POST-RACE'? BACKGROUND AND PRECAUTIONS

When I tell my colleagues, friends and other conference attendees that I am looking into post-race theory in educational research, I am generally met with two kids of reactions: '*Post-race? I thought we were already!*' and '*How can you say race doesn't matter anymore? It's important to people!*' These comments are understandable considering how the term 'post-race' remains highly ambiguous. The phrase 'post-race' can be perceived as suggesting that our societies successfully eliminated racism, or racism now plays little significance in people's lives. Neither of these cases is true; it would be downright wrong, and dangerous, to say that we now live in a raceless world. The literature on colour-blind racism demonstrates how hasty adaptation of 'raceless' perspective can lead to reification of persisting inequalities between historically demarcated groups (Bonilla-Silva, 2003; Zuberi & Bonilla-Silva, 2008). Likewise, Warmington (2009) contends that 'race may lack scientific integrity but it is a lived experience, a lived relationship' (p. 283), and that '[w]e are post-racial in having moved beyond pseudo-genetic notions of race; however, we are not post-racial *per se*' (p. 295).

While 'race' continues to be significant, we now recognise a vast complexity and multiplicity in ways that people's 'race' is talked about, observed, enacted,

M. Vicars et al. (eds.), Discourse, Power, and Resistance Down Under, 95–106.

imposed and questioned. Accordingly, it is simply not possible to speak of such things as *the* racial minority experience or perspective. Varying degrees of ethnocentrism and racism are expressed within communities, and it is clear that exclusivist and separatist ideologies are not leveraged solely by 'whites.' Just as much as some (self-defined) 'whites' demonstrate strong tribalism and prefer to 'stick to their own kinds' and participate in discriminatory practices towards out-groups, so do some members of historically marginalised groups. Some 'minorities' wish they could transcend existing categorical boundaries and do not constantly need to define themselves (or be identified) on the basis of affiliations based on 'race,' even though this could mean forfeiting access to substantial benefits. Others, however, feel that the sense of belonging and collective power founded on racial inclusion (and exclusion) are so important that the idea of 'race' cannot be lightly given up (Gilroy, 2000).

Another recent change that we see is how historically distinct communities are *transfusing* due to various processes including increased global mobility, inter-marriages and development of information and communication technologies. Usage of words like 'race,' 'ethnicity' and 'culture' can no longer be limited to the nation-centric manner; the term 'Australian' no longer has the same degree of racial ('Anglo' or 'White') connotation as it once did. In some contexts previously held 'racial' boundaries between groups are becoming weakened, while in other contexts they have been newly constructed or reconstructed due to changes in political and economical situations. I have for the last few years been witnessing a fast-paced resurgence and intensification of Japanese nationalism and racism – reminiscent of pre-WWII imperialism – as a result of economic advancements of South Korea and China. Ideas about social identities, informed by 'race,' ethnicity and culture are constantly transformed, mobilised and negotiated in response to a range of local, trans-local and global processes. It is within this space that I consider the possibilities of post-race theory.

While the idealistic outlook of 'post-race' can be appealing, there is a danger that such perspective is prematurely adapted, or worse, lend itself to naïve liberalism which perpetuates racial discrimination. Leonardo (2005) reminds us that it is impossible for us to be *ante*-race as much as we can be *against* or *anti*-race. The idea that 'race' as a necessarily and legitimate means of social categorisation is so deeply woven into the way people behave and communicate, and is readily transmitted to younger generations of and newcomers to our societies. Young children in preschool and primary schools have been observed to adapt and utilise the discourse of race and racial hierarchy in sophisticated manner, moderating power relationships and group memberships between peers (Ali, 2003; Rossholt, 2006; Van Ausdale & Feagin, 2001). The utmost dilemma that we face today is that, although the biological argument for racial differences has been seriously undermined (Zack, 2001), racial inequality is so deeply ingrained in our social structure that the deliberate 'colorblindness' analysis cannot be considered as a remedy. In fact, it has been argued by many that universalist, one-size-fits-all practice based on the principle of formal equality will merely sustain the existing inequality (see e.g., Arthur, 2007; Gilborn, 2008).

For these reasons, I must stress that post-race theory is a tentative and experimental project, and that it is not intended to construct a 'new' grand narrative in thinking about 'race.' Another reason I do this is because of the tendency in my (educational) discipline that I have witnessed in my candidature. Surprisingly many people see theories as prescriptive and normative, that is, that theories define how one *should* think. While some theories such as feminist theories and Critical Race Theory sometimes (self-consciously) present themselves as having clear ethical and political commitments, I maintain that theories are fundamentally impartial (van den Berghe, 1986) and their function is to enable us to reduce and simplify the complex social reality in particular manners, so that we can have a shared understanding and discourses. Furthermore, people often seem to have a competitive view of theory, that is, they see it as though I am attempting to replace existing frameworks with post-race theory, or to show that post-race theory is inherently and universally superior. *No* proper theory is (and historically speaking, was) universally accurate or 'right' (Popper, 1963; Kuhn, 1962). Even the natural science has been criticised as being partial and imperialist, and does not necessarily align with perspectives held by Indigenous groups around the world. Accordingly, I see that post-race theory is necessarily constrained, like any other existing and well-regarded frameworks. Its virtue is that it allows us to capture and deal with different types of experiences and accounts of 'race' which other frameworks may not. Rather than being competitive, post-race theory is both *complement* to existing theoretical frameworks, with overlaps and contradictions; yet it shares the same goal of deepening understanding and knowledge of "the *continuing significance and changing meaning of race*" (Winant, 1993/2000, p. 182; emphasis original).

THE LITERATURE ON POST-RACE THEORY

The amount of literature explicitly dealing with post-race theory is limited at this point in time. Within the small body of the literature I managed to identify (Ali, 2003; Byrne, 2011; Nayak, 2006; Warmington, 2009; Zack, 2007), references are often made to Paul Gilroy's work. Gilroy, in his 1998 article *Race ends here* and in 2000 book *Between camps* (published as *Against race* in the United States) discusses what he calls the crisis of 'race' and raciology. Raciology is a term Gilroy used to refer to 'discourses of the race-difference and all the stereotypes, prejudices, images, identities and knowledges it carries in its wake' (During, 1999, p. 264). Gilroy (2000) claimed:

> The pursuit of liberation from "race" is an especially urgent matter for those people who, like modern blacks in the period after transatlantic slavery, were assigned an inferior position in the enduring hierarchies that raciology creates. … Black and white are bonded together by the mechanisms of "race" that estrange them from each other and amputate their common humanity. (p. 15)

Suki Ali (2003), who conducted a study on 'mixed-race' primary school children in the UK, agrees with Gilroy's view that 'we need "to free ourselves from the bonds of raciology" and "compulsory raciality"' (p. 18). Ali's description of her post-race project is as follows: 'My use of the term 'post-race' thinking emphasises deconstructive approaches to identities, and draws on theories of performativity, passing and new ethnicities' (p. 9). The thesis that 'race as performativity' – adaptation of Judith Butler's work on the construction and reification of 'gender' identities (1990) – appears to be one of the underpinning themes of post-race theorising. In her book *Gender Trouble,* Butler described 'gender' as 'the repeated stylization of the body, a set of *repeated acts within a highly rigid regulatory frame* that congeal over time to produce the appearance of substance, of a natural sort of being' (p. 33; emphasis added). This view is particularly compelling in societies like Australia, the United Kingdom and the United States. The number of 'mixed-race' and transnational children are rapidly increasing (Dewan, 2008), and more people are having close affiliations (ancestral or otherwise) with people in places outside the communities and countries where they were born or currently live (Rizvi & Lingard, 2010; Weedon, 2004). For these people, racial and cultural identities are not necessarily attached to a singular geographical location, or a particular group that they were born into. The emerging literature on 'affiliative' ethnic identity also makes this point. Affiliative ethnic identity depends on 'knowledge, consumption and deployment of ethnically linked symbols and practices' (Jiménez, 2010, p. 1757) rather than claims of ancestry. As Butler points out, however, 'gender' (or 'race') as performativity does not mean that we have free choice; the kind of choice that one can have, and what acts count as constituting particular identity is constrained by 'highly rigid regulatory frame.' Evidently, Ali (2003) finds in her studies that mixed-race children are often exposed to and are resistant to 'the dominant discourses of 'race' that forced them to choose singular positions' (p. 180), and they can 'feel obliged to move into a position of singularity, at least for 'political' purposes' (ibid.). Similarly, when people seek affiliative identities, it is easy to imagine that their performances would be necessarily (and deliberately) constrained by the dominant discourse which prescribe readily recognisable behaviours and practices of the desired groups (e.g., clothing, music, speech style). On the one hand this performance does disrupt the discourse of singularity, yet on the other hand it merely reinforces the 'authenticity' based on stereotypical and essentialist views.

While Ali's work (2003) mainly concerned the ways 'race' is constituted and performed in everyday schooling context, Anoop Nayak (2006) focuses on how 'race' is *re*produced via researchers' ethnographic gaze. He argues that the popularity of poststructuralist and social constructionist critique of 'race' is 'yet to halt the dense economy of signs and signifiers that proliferate in contemporary' (p. 412). Dewan (2008) also claimed that the critique of race as a mere social construction has by now 'somewhat hackneyed in the academic field' (p. 8) yet such academic move has done a little to challenge the foundations of raciology.

Underpinning Nayak's post-race proposal is Jacques Derrida's work on semiotics. For Derrida, 'race' is a signifier which is reiterated through the act of repetition; the semiotic relationship between the terminology 'race' and what we recognise as 'race' is contingent upon our repeating it (or not). Using the term 'race,' or even 'post-race,' in discussions and conversations presupposes certain intersubjective understanding of what 'race' entails. O'Brien's (2011) shares a similar concern, claiming that race-conscious research data and analysis can make 'race and ethnicity seem like essentialized qualities of human bodies' (p. 69). O'Brien emphasises that 'race' is 'activated' in research and every practices:

> we need an approach to in-depth interviewing that does not reduce "race" and "ethnicity" to formulaic categories, but approaches them as ways of relating to each other that vary not just by phenotype, but by *the way each of us "activates" race (or not) in our everyday conversations and lives.* (p. 67; my emphasis)

How, then, could we – as paradoxical it may sound – try to write race into absence? Nayak's (2006) proposition is that post-race writing can challenge the preconceived ontological and epistemological status of racial identity 'by adopting an anti-foundationalist perspective which claims that race is a fiction only ever given substance to through the illusion of performance, action and utterance, where repetition makes it appear as-if-real' (p. 416). Nayak goes on to argue that

> The radical potential in this [post-race] perspective lies in the understanding that our cultural identities are *produced* in the ethnographic encounter rather than coming to precede the event. They go beyond the now familiar post-structuralist notion of in-depth ethnography as simply power-ridden terrain where multiple subjectivities of race, class, gender and sexuality intersect. Instead of seeing race as a dimension we bring to the interviewing table, a post-race reading would stress the *impossibility* of this identity (pp. 425-426)

The point that identity is *produced* in the ethnographic encounter, rather than preceding it, is highly relevant to current debates on voice, subjectivity and narrative inquiry. For example, White and Drew (2011) warn against the romanticised notion of voices. Voices are elicited through interviews and narratives in a highly contextualised and specific environment, and it can be dangerous to accept these voices as true and holistic representation of experience of interviewees and narrators. Traditional assumption about researchers' subjective neutrality has been under question, too (see Chase, 2010; Henry, 2010; Luttrell, 2010). Researchers inevitably shape the nature of ethnographic encounter to elicit particular kinds of voices and performances. In this perspective, it can be considered that 'racial' identities and 'racialised' voice of narrators and research participants are, to an extent, artefacts of research design and process.

Nayak's (2006) post-race outlook is applied to an Australian educational context by Zink (2007). Zink interrogates the construction of 'Indigenous good / non-Indigenous bad' racialised binary in Victoria state's Outdoor Environmental Studies curriculum. It is argued here that the reification of Indigenous Australians as being 'ecologically noble (= good)' is problematic, as

> Constituting Indigenous as good only makes sense in relation to non-Indigenous as bad. Both who and what counts as good and bad changes

> depending on the needs of the social and cultural context. Indigenous are cast as bad if there is a need to posses land or enhance the notion of civilization of Western society. Alternatively, Indigenous are cast as good when aspects of Western society are out of favour. (p. 5)

I read Zink's concern as depicting how contemporary politics in Australia not only legitimatise a particular means of 'racial' and 'ethnic' demarcation but also idealise the notion of historically disparate identities. While rendering Indigenous peoples in a positive light is commendable, it does little to challenge the conceptual centrality of 'white' category, a reference point from which all the Others have been measured up and marginalised. As Zink notes, such dichotomised representation of Indigenous and non-Indigenous peoples 'freezes race as fixed and essential' (p. 7), reinforcing the legitimacy of particular discourses about 'race.' Gilroy (2000) also raised a similar point concerning the recent transformation of 'blackness' from a badge of insult to the sign of beauty. While the deliberate celebration and appreciation of 'blackness' may appear pleasing, one may be sceptical of this shift as it is driven by corporate interests that attempt to promote contrived notions of racial authenticity, inclusivity and exclusivity. He maintains:

> It is best to be absolutely clear that the ubiquity and prominence currently accorded to exceptionally beautiful and glamorous but nonetheless racialized bodies do nothing to change the everyday forms of racial hierarchy. The historic associations of blackness with infrahumanity, brutality, crime, idleness, excessive threatening fertility, and so on remain undisturbed. (p. 22)

In a similar vein, the discourse of 'Indigenous good / non-Indigenous bad' can still subordinate the Indigenous peoples, as the value of (unitarily represented) 'Indigenous' identity, culture and knowledge are determined in accordance to external political interests. Another underlining issue is that such discourse inevitably reiterates the quasi-polygenist or -evolutionist account of 'race,' where each group is ranked as essentially more capable or suitable than others for specific purposes. The only difference to old discourses of 'race' (Spencer, 2006) is that we no longer assume that 'whites' are equivocally superior over the other group. It is still assumed that 'whites,' 'blacks,' 'Hispanics,' 'Asian' and so forth are essentially different from each other.

To summarise, my reading of the small body of available literature on post-race theory provides several important characteristics of post-race theory that are worth consideration. I will, however, limit my discussion in the reminder of this chapter to one of such characteristics. I will supplement and expand on what Nayak (2006) calls anti-foundationalist critique of 'race.' I will also demonstrate how the development of post-race theory can be situated in the contemporary work of global connectivity, global consciousness and cosmopolitan learning.

ANTI-FOUNDATIONALIST CRITIQUE OF RACE

Anti-foundationalism (or antifoundationalism) is a doctrine which rejects the idea that there is a single foundation through which every piece of knowledge is ultimately interrelated. What we call 'knowledge' would not exist outside our consciousness; constitution of knowledge relies on shared belief and consensus. In this view, racial and cultural identities are what *we* continuously reproduce and agree upon. I can only *be* as 'Asian,' 'Australian' or 'Japanese' as a particular intersubjective space permits me to be. It is 'impossible,' as Nayak (2006) puts it, for me to concretely achieve such identity because there is nothing outside of our own will that necessitates or validates it. To say that we recognise 'racial' and 'cultural' identity, difference and diversity can tell us about the kind of cultural conventions that shape our thinking (Nagami, 1985).

It was during the 1980s when cultural and semiotic analysis of race started to emerge, although one may trace this line of work to a small number of scholars from much earlier such as W. E. B. Du Bois and Franz Boas (Winant, 2000). Stuart Hall (1981, 1997) has discussed the conceptual instability of racial and cultural identities, and rendered 'race' as a *floating signifier* whose utility of demarcation upon historically specific conditions. Omi and Winant (1986) rendered 'whiteness' (or any 'racial' identity for that matter) as a category only given substance and understood through the construction of 'black' Other. These critiques of race – and emerging work of post-race theory as discussed here – are not to (nor can they) override other ways of theorising racial and cultural identities, or undermine specificities of existing groups. Evidently we do enjoy and benefit from unique philosophies, artefacts, technologies and skills that have been cultivated by historically and geographically demarcated groups. 'Multicultural' events and festivals, no matter how contrived and superficial, can provide rare opportunities for some people to interact with and learn about other traditions and cultural conventions than their own; or to recognise that their culture, too, can appear different and peculiar from others' perspectives. The 'good' (however one defines it) can comes out of collective actions and interests of people who are bounded by so-called 'race' and 'ethnicity,' and identifications with such groups can provide people with collective power and sense of belonging that may be beneficial in some situations. 'Racial' categories can be considered as necessary for the protection and empowerment of historically disadvantaged and vulnerable groups of people (e.g., affirmative actions).

Instead of refuting the instrumental value of racial categories *per se*, what anti-foundationalism can offer is a precaution to how the rigidity and fixity of 'race' concept is often taken for granted in research practice. Today, we can find a myriad of confirming evidence that tells us how race is important – if we look for them, just as we can do the contrary by selectively focusing on events that tell us how celebration of racial diversity through simplistic institutional categorisation and labelling can be dangerous and detrimental (see, e.g., Arnesen, 2010). Anti-foundationalist outlook is beneficial here as it clearly draws the distinction between two classes of 'knowledge': *solidarity* and the *objectivity*. The number of confirming evidence that 'race' is important and make it seem that 'race' is absolutely an essential concept. But this is no more than a solidarity, a shared belief or consensus within a community, rather than objectivity (Rorty, 1991). What this means is that *we* simply believe in the existence of particular kinds of 'racial' and 'cultural' identities because *we* feel it is beneficial for *us* to do so. What constitutes 'beneficial' is obviously largely dependent on who *we* are (researchers? Teachers? Policy makers? Members of oppressive minority? Bigots?), what *we* do and for what purpose (theoretically understanding the complexity of domination and interrogating it? Provision of remedial treatments? Preserving sense of group solidarity by including and excluding particular individuals?). Whatever our circumstances are, as we see in the formal logical fallacy called 'affirming the consequent,' outcomes do not justify its premises; we can never induce the objective status of 'race' by observing the current social and political situations. In this sense, 'post-racial' criticism of 'race' from anti-foundationalist standpoint provides us researchers a precautions to qualitative researchers about how our 'knowledge' may be limited to solidarity based on cultural and historical conditions; and that 'race' have become so deeply entrenched in our institutional and political fabric that they are given a quasi-objective status in everyday discourses.

Another work that can aid us to understand post-race critique of 'race' is Foucault's work of critical ontology. Critical ontology was presented under three headings: 'How are we constituted as subjects of our own knowledge? How are we constituted as subjects who exercise or submit to power relations? How are we constituted as moral subjects of our own actions?' (Foucault, 1984, p. 49). These questions help us 'to examine ideas and principles that organize our habitual ways of thinking and acting in order to think and act differently' (Wong, 2007, p. 73). Instead of rejecting everything, critical ontology is a practical critiques of existing knowledge system to allow possible transformations. It takes an experimental attitude towards existing social and political reality, 'both to grasp the points where change is possible and desirable, and to determine the precise form this change should take' (Foucault, 1984, p. 46).

POST-RACE THEORY, GLOBAL CONSCIOUSNESS AND REFLEXIVITY

Since studies of 'race' deal with topics that are highly contingent, we must also consider how post-race theory can inform, and is informed by, current social changes. Among many chances that are taking place, I find that the influences of globalisation has been especially pertinent in various research communities. Rizvi and Lingard (2010) describe the situation as follows:

> Under the condition of globalization, the assumption of discrete national cultural formations can no longer be taken for granted, as there is now an ever-increasing level of cultural interactions across national and ethnic communities. With the sheer scale, intensity, speed and volume of global cultural communication, the traditional link between territory and social

> identity is now much more complicated; people now more readily choose to detach their identities from particular times, places and traditions. (Rizvi & Lingard, 2010, p. 30)

The global media and greater transnational mobility has led to formation of less 'fixed or unified,' hybrid identities (Rizvi & Lingard, 2010). This transformation does not only apply to the people that we observe and write about. More researchers today see themselves as having such identities as they travel between countries; they also develop a greater global consciousness through which they locate themselves and understand the embodiment of specific cultural and political values and presumptions (see, e.g., Henry, 2010; Pemunta, 2010). Stanfield (2011) documents how his own development of global consciousness and lifestyle has had influences on his work as a sociologist. Reflecting on the earlier version of the volume *Rethinking race and ethnicity in research methods* which was published in 1993, he admits he 'was entrapped by [his] own nationalistic ethnocentrism' (p. 14). He also reflects on how he had 'lacked a global sense of consciousness and identity that rendered [his] international interests theoretical and quite secondary to [his] nascent Americanism' (p. 15).

Global consciousness also allowed Stanfield to critique how some researchers tend to romanticise communities and peoples that they study about. Romanticisation of culture, tradition, heritage and so called 'roots' can be dangerous for researchers (and readers) because it can make these social factors appear to be more salient. Gilroy (2000) sees that multiple axes of differences within communities can be used to 'challenge the unanimity of racialized collectivities' (p. 24). However, once the social imaginary of racialised collectivism is romanticised by the researcher, it is easy to see that complex differences that people possess become supplanted by the notions of shared past and history, further reinforcing researchers' desire to see the world through 'racialised' and 'culturalised' fashion. Problems of such epistemological preoccupation is outlined by Nagami (1985):

> 1) Those who accept the group versus individual thesis [e.g., that Japanese in general behave in group oriented manner while Americans behave on the basis of individual freedom] tend to explain every cultural phenomenon in terms of this and as a result they conceal many other rich possibilities which every culture may contain. 2) Certain features are common to all social worlds. In this respect every culture can share a kind of symbolic common denominator through which we can compare differences. Yet, this approach of contrast ignores the universality in human existence. 3) The very fact that people understand and describe different cultures reveals a kind of cultural ethos in which their thinking is embedded. That is to say, their ways of thinking are inescapably and necessarily cultural. (cited in Solomon, 2001, p. 305)

The significance of such self-reflexive approach to inquiry is elucidated by Rizvi (2006). Rizvi discusses how cosmopolitan learning should be aimed at developing alternative imaginary of global interconnectivity

> which is informed not by the universalizing logic of the market, or by the romanticised notions of global citizenship, but by our determination to develop a different conception of global relations, which views all of the world's diverse people and communities as part of the same moral universe. (pp. 31-32)

Such imaginary, Rizvi contends, is coupled with the development of sets of epistemic virtues. One of such virtues is reflexivity, enabling 'a critical recognition of our own cultural and political presuppositions' (p. 33). Reflexivity allows individuals to 'challenge their own taken for granted assumptions that are often linked to official and popular discourses of cultural difference' (ibid.) and 'reflect upon the politics of their own representations of others, and point to the ways in which this politics is historically constituted' (ibid.). Likewise post-race theory, as I have try to demonstrate here, can allow us to reflect on the way we have learned to theorise and conceptualise 'race,' and provide new understanding and questions about such topics as social relations, diversity, hybridity inclusivity and multiculturalism.

SUMMARY

In this chapter, I have introduced a newly emerging and highly contentious theoretical perspective called 'post-race theory,' and explored one of its tenets: anti-foundationalist critique of 'race' concept. My argument is that the virtue of post-race theory is the reflexive gaze that it requires researchers to adapt when thinking and writing about 'race.' I have also argued that, given the vast complexity and controversies surrounding the conception of 'race,' post-race theory should be understood as complimentary to, rather than in competition with or attempting to supersede, existing theoretical frameworks. Overall, my intention has been to open a critical discussion and to demonstrate how post-race theory may be an exciting intellectual position for researchers to consider.

REFERENCES

Ali, S. (2003). *Mixed-race, post-race: Gender, new ethnicities and cultural practices*. Oxford: Berg.

Arnesen, A.-L. (2010). The complexity of diversity – Challenges for teacher education. In *Policies and practices for teaching and sociocultural diversity* (pp. 21-34). Strasbourg Cedex: Coucil of Europe Publishing.

Arthur, J. (2007). *Race, equality, and the burdens of history*. Cambridge: Cambridge University Press.

Bonilla-Silva, E. (2003). *Racism without racists: Color-blind racism and the persistence of racial inequality in the United States*. Lanham: Rowman & Littlefield.

Butler, J. (1990). *Gender trouble: Feminism and the subversion of identity*. London: Routledge.

Byrne, B. (2011). Post-race? Nation, inheritance and the contradictory performativity of race in Barack Obama's 'A More Perfect Union' speech. *Thirdspace: A Journal of Feminist Theory & Culture, 10*(1). Retrieved from http://www.thirdspace.ca/journal/article/view/byrne.

Chase, S.E. (2010). Narrative inquiry: Multiple lenses, approaches, voices. In W. Luttrell (Ed.), *Qualitative educational research: Readings in reflexive methodology and transformative practice* (pp. 208-236). New York: Routledge.

Dewan, I.A. (2008). *Recasting race: Women of mixed heritage in further education*. Stoke on Trent, UK: Trentham Books.

During, S. (Ed.). (1999). *The cultural studies reader*. London: Routledge.

Foucault, M. (1984). What is enlightenment? In P. Rainbow (Ed.), *The Foucault reader* (pp. 32-50). New York: Pantheon Books.

Gillborn, D. (2008). *Racism and education: Coincidence or conspiracy?* New York: Routledge.

Gilroy, P. (2000). *Against race: imagining political culture beyond the color line*. Cambridge, MA: Belknap Press of Harvard University Press.

Henry, M.G. (2010). "Where are you really from?": Representation, identity and power in the fieldwork experiences of a South Asian diasporic. In W. Luttrell (Ed.), *Qualitative educational research: Readings in reflexive methodology and transformative practice* (pp. 363-373). New York: Routledge.

Jiménez, T.R. (2010). Affiliative ethnic identity: A more elastic link between ethnic ancestry and culture. *Ethnic and Racial Studies, 33*(10), 1756-1775.

Leonardo, Z. (2004). The color of supremacy: Beyond the discourse of 'white privilege' *Educational Philosophy and Theory, 36*(2), 137-152.

Luttrell, W. (2010). "Good enough" methods of life-story analysis. In W. Luttrell (Ed.), *Qualitative educational research: Readings in reflexive mthodology and transformative practice* (pp. 258-278). New York: Routledge.

Nagami, I. (1985). Cultural gaps: Why do we misunderstand? In R.C. Solomon (Ed.), *Introducing philosophy: A text with integrated readings* (7th ed., pp. 304-310). Oxford: Oxford University Press.

Nayak, A. (2006). After race: ethnography, race and post-race theory. *Ethnic and Racial Studies, 29*(3), 411-430.

O'Brien, E. (2011). The transformation of the role of "race" in the qualitative interview: Not if race matters, but how? In J.H. Stanfield, II. (Ed.), *Rethinking race and ethnicity in research methods* (pp. 67-93). Walnut Creek, CA: Left Coast Press.

Pemunta, N. V. (2010). Intersubjectivity and power in ethnographic research. *Qualitative Research Journal, 10*(2), 3-19.

Rizvi, F. (2006). Epistemic virtues and cosmopolitan learning.Unpublished manuscript.

Rizvi, F., & Lingard, B. (2010). *Globalizing education policy*. New York: Routledge.

Rorty, R. (1990). *Objectivity, relativism, and truth: Philosophical papers, Volume I*. Cambridge: Cambridge University Press.

Rossholt, N. (2006). Inscribing the body: 'Black,' 'white' and gender in early childhood education and complexity. *Journal of Australian Research in Early Childhood Education, 13*(1), 112-124.

Smith, L.T. (1999). *Decolonizing methodologies: Research and indigenous peoples*. New York: Pelgrave.

Spencer, S. (2006). *Race and ethnicity: Culture, identity and representation*. London: Routledge.

Stanfield, J.H., II. (2011). Epistemological reconsiderations and new considerations: Or what have I been learning since 1993. In J.H. Stanfield, II. (Ed.), *Rethinking race and ethnicity in research methods* (pp. 11-26). Walnut Creek, CA: Left Coast Press.

Van Ausdale, D., & Feagin, J.R. (2001). *The first R: How children learn race and racism*. Maryland: Rowman & Littlefield Publishers.

van den Berghe, P.L. (1986). Ethnicity and the socio-biology debate. In *Theories of race and ethnic relations* (pp. 246-263). New York: Press Syndicate of the University of Cambridge.

Warmington, P. (2009). Taking race out of scare quotes: race-conscious social analysis in an ostensibly post-racial world. *Race Ethnicity and Education, 12*(3), 281-296.

Weedon, C. (2004). *Identity and culture: Narratives of difference and belonging*. London: Open University Press.

White, J., & Drew, S. (2011). Collecting data or creating meaning? Troubling authenticity in ethnographic research. *Qualitative Research Journal, 11*(1), 3-12.

Winant, H. (2000). The theoretical status of the concept of race: A reader. In L. Back & J. Solomos (Eds.), *Theories of race and racism* (pp. 181-194). London: Routledge.

Wong, J. (2007). Paradox of capacity and power: Critical ontology and the developmental model of childhood. In M.A. Peters & T. Besley (Eds.), *Why Foucault? New directions in educational research*. New York: Peter Lang.

Zack, N. (2001). Philosophical aspects of the 'AAA Statement on "Race."' *Anthropological Theory, 1*(4), 445-465.

Zink, R. (2007). Can we move beyond 'Indigenous good, non-Indigenous bad' in thinking about people and the environment? *Australian Journal of Outdoor Education, 11*(2), 3-9.

Zuberi, T., & Bonilla-Silva, E. (2008). *White logic, white methods: Racism and methodology*. Boulder, CO: Rowman & Littlefield.

Satoshi Sanada
School of Education
James Cook University, Australia

STEFAN SCHUTT AND MARSHA BERRY

HOW THE 'I' SEES IT

The maker, the researcher and the subject at the juncture of memory and history

INTRODUCTION

In this chapter we attempt to come to terms with our simultaneous roles as researchers and research subjects who use creative practice (primarily video, text and installation) to shed light on our own postmemorial urges (Hirsch, 1997). Our raw materials are our personal encounters with the traumatic ghosts of an intimate, yet distant, past: those of family stories preceding our birth. Woven into this exploration are issues connected with the effects of modernity and our digital era in particular, especially the ever-expanding range of media creation tools for resuscitating and disseminating aspects of the past that, once revived, haunt the present (Derrida, 1993; Gordon, 1997). We explore the ways in which these methods have simultaneously fuelled and complicated what we do, how we do it, and how this relates to others' practices of remembering. This chapter asks: where is the researcher's place when working with one's own traumatic ghosts, and where are the fault lines between the self, the use of creative practice as a mode of tacit knowledge production, and the role of 'researcher?'

HISTORY, MEMORY AND POSTMEMORY

We begin this exploration with historian Pierre Nora. In *Les Lieux de Mémoire* (Sites of Memory) Nora points to a widening rupture between memory and history, fuelled by 'our hopelessly forgetful modern societies, propelled by change.' Memory, states Nora, 'is life ... it remains in permanent evolution' whereas, 'history, on the other hand, is the reconstruction, always problematic and incomplete, of what is no longer' (1989, p. 8).

The "I" sits at the juncture of memory and history, witnessing and recording, turning the lived experience of memory into history. The tools of the digital age have made it easier than ever for many people to assume the mantle of this 'I,' not just the professionals of historiography referred to by Nora in his pre-Internet analysis. As we write this, thousands are busily disinterring and resuscitating or reconstituting the forgotten and the buried, placing the resulting media artefacts – images, videos, stories, dates, names, sounds – in online public spaces where other 'I's can search for and find them, then use them as raw material for their own reconstructions. We see this in an increasing range of software-facilitated forms: the scanning and uploading of old photos, the compiling of personal videos, the

M. Vicars et al. (eds.), Discourse, Power, and Resistance Down Under, 107–119.

creation and sharing of online family histories and genealogies, the simulation of 1970s Polaroids and Super 8 movie footage. We live in an era of Derrida's *hauntology* (1993), when masses of ghosts are being conjured up, unsettling an uncertain present. And it is we who are generating these ghosts, a society under the grip of nostalgia in its literal sense of an amalgam of the Greek words for 'returning home' and 'pain' or 'ache': literally, 'homesickness,' a grasping for something lost.

As creative practitioners and researchers, we too make use of digital tools to explore and share what it means to remember. Our subject matter consists of our own family ghosts: our respective forebears and their experiences of displacement and trauma. We seek to encounter and converse with these ghosts, and through our encounters, interpretations and reinterpretations hope to develop an aesthetic of memories, postmemories and auras.

The raw materials we use include family photographs that have literally haunted us. Figure 1 shows a family bound for Berlin from Soviet Russia in 1933. They are thin and their faces are etched with their struggles for survival.

Figure 1. Family photograph, USSR, 1933

The image takes central place in the family photo albums now owned by the descendants of the faces in the photograph. Digitised and copied, it is easily shared amongst a family scattered across the globe, a family that is still affected by the events it represents. Of the people recorded at this particular, pivotal moment in their family's history, Natasha (bottom right) was the only child excited by the family's imminent emigration. She had been to Berlin for a sojourn with her mother in 1930 while her father was still incarcerated in the Solovetsky Islands (Gulags) under Sections 58.10 to 58.11 for counter-revolutionary activities. He spent four years in the Gulags and was released shortly before this photograph was taken.

In examining this photograph we note its aura, referred to by Walter Benjamin in *A Short History of Photography* as a 'strange weave of space and time: the unique appearance or semblance of distance, no matter how close it may be' (1932, p. 119). As Briggs (2008) points out, however, 'the aura is neither a stable attribute nor an object, but an index of the dynamic fraught relationship between the beholder and the artefact' (p. 115). Indeed, looking back now at this image, we might add to the 'strange weave of time and space' a kind of inverse impression – a semblance of closeness, no matter how temporally distant the events may be now.

The photograph below was found in a shoebox of images belonging to Stefan Schutt's grandfather and discovered in Germany in 2005. It depicts Schutt's grandmother's family estate in Pomerania in 1944, months before the arrival of the Russian front and all its privations, and the subsequent expulsion of the German population and demolition of the estate house. Figure 2 is a digital photograph Schutt took of the original photograph during a field trip.

Figure 2. Family photograph, Koldemanz, 1944

The power of this photograph to haunt has previously been analysed by Schutt and Berry (2011) as one connected to its *punctum*: the hidden, wounding, subterranean counter-meaning serving to puncture the studium, its ostensible surface meaning (Barthes, 1980). Like the haunting spectres that invade the present, the punctum is slippery; it 'eludes, evades, escapes our efforts to ensnare it and catch it in meaning. It gives sense without being sensible' (Chare, 2008, p. 96). This forces us to look outside the frames we have created: 'The punctum is not part of the meanings that we give to an image, rather it gives us meaning. Through the injury it causes, I feel something and I must ask why have I felt and what have I felt' (p. 96). In both photographs, the injuries they impart are highly contextual and depend upon a sense of family connection for their power. In looking at this more closely, we have drawn a connection between this power and the theory of postmemory, upon which we will expand shortly.

The photographs above also hint at the importance of place to our work. We have travelled to some of the important locations featured in our family stories, digitally recording our experiences through video, images, audio and text to generate raw material. These places were the settings for significant events in our family histories that occurred before we were born. The backgrounds and experiences of our respective families differ considerably yet they are also linked through place and time – World War II Germany – and the effects of war and political upheaval.

Our personal fieldwork takes its cue from the life mapping undertaken by Walter Benjamin in *Berlin Childhood Around 1900*, with its confluence of geography and auto/biography, and its sense of an archaeology of the self. But the two also differ. Benjamin's series of vignettes (thought pieces, or '*denkbilder*') depicted childhood life in a middle-class, turn-of-the-century German-Jewish household, and was a form of inoculation against future exile and displacement (1932b, p. viii), an attempt to come to terms with what Benjamin first sensed, then knew, was to come (the work was written and revised both before and after the Nazis' rise to power). Our field trips were an attempt to come to terms with past family exile and displacement, as experienced vicariously a generation later. For Benjamin and for us, particular locations possess the power to trigger profoundly strong and mixed emotions, a kind of haunting that troubles the present. In Benjamin's case, these locations were familiar: inhabited, remembered, then recalled in his writing. Our locations were unknown yet not entirely unfamiliar. By visiting them we felt unspoken connections to an ancestral past, connections with origins in family photographs, home movies and the stories that went with them. Whereas Benjamin was haunted by an unfocused but pervasive sense of future loss, we were haunted by an unfocused but pervasive sense of past loss.

Two more examples follow. The first, an extract from Marsha Berry's travel journal, reads:

> Dresden, April 6, 2007.
>
> My parents migrated to Australia in 1950 as Displaced Persons from a German refugee camp. Both were former Soviet citizens fleeing Stalinist policies. As a small child living in Melbourne's Surrey Hills, I heard many conversations my mother, Natasha (1920-1998) had with my father (1919-1988) reminiscing about their war experiences.
>
> Mum's stories were about Berlin and Dresden. In 1944 she was living in Dresden. She left in the morning (Feb 14) of the night of the first carpet bombings. She watched the planes fly over in the evening and then saw an orange glow on the horizon. She realised Dresden was burning. She found the sight incredibly beautiful yet at the same time she could picture the horror in

her mind. She always said that her guardian angel had spoken to her, urging her to leave. I was haunted by her story. It had become part of me.

I am overwhelmed by Being Here today in Dresden. I feel like I am on a pilgrimage. When we drove into the city the sight of the heart of Dresden restored is impossible to capture. It is so achingly beautiful, sublime, I want to vanish here, to be part of the cobblestones. I find it hard to believe I am actually here … so many feelings: loss, belatedness, being born too late … I feel so connected to this place, I am not just a tourist of traveller like I was in Leipzig, somehow I feel part of me belongs here even though this is impossible.

S [cousin] and I walked the streets of the restored city together. We were overwhelmed see Frauenkirche restored to its Baroque glory. D [L, another cousin's son] apartment is on the top floor of an apartment building – early 19th century – by coincidence it fits the description of the apartment Mum had lived in. I can almost see her here.

I took so many photos and video tonight. I am looking forward to taking lots more tomorrow. Now it's time for me to make memories of Dresden, of my Dresden.

Next we present a photograph taken by Schutt during a 2005 visit to the site of the estate shown in the previous photograph. This was the first visit by a family member to the village of Kolomac, Poland (previously Koldemanz, Germany) since the fall of the Iron Curtain. The photograph shows the ruins of the church where Schutt's grandparents were married in 1937. The church stood next to the family estate house depicted in the previous photograph. As was the case in villages throughout Pomerania after World War II, all reminders of the area's former German population were demolished including the church, the estate house and the cemetery attached to the church.

Figure 3. Church ruins, Kolomac, 2005

Our stories are particular to our family histories, but our experiences of family pilgrimage are by no means unique. Children of displaced or exiled survivors of trauma grow up with their parents' stories of places, people and events, and take them on as part of their own identities. Media objects such as family photos and films are often associated with such stories. Marianne Hirsch (2001) has coined the term *postmemory* to describe and theorise this phenomenon, which emerged out of Hirsch's work with the children of Holocaust survivors.

In postmemory, stories and images of life before and during momentous or traumatic events become ingrained in the early childhood of the next generation as a pervasive background against which they interpret their lived experience. This can sometimes overshadow children's lives, in that the lives of the survivors may feel more significant or epic. Postmemory is temporally and spatially different from survivor memory. It is mediated not through direct remembering but through imagination and empathy, displacing the here and now.

There are many poignant accounts of children of survivors visiting places where their parents had lived before the Holocaust, using photographs as clues to find exact locations, and filling in the gaps through imagination, in an attempt to understand what their parents went through. Such visits are often accompanied by a feeling of familiarity, belatedness and a sense of absence or gaps as well as a quest for emotional truth. Hirsch herself experienced this. Her and Leo Spitzer's (2006) account of their return to Czernowitz, now Chernivtsi, in the Ukraine is peppered with a need to find emotional truth as well as numerous references to their sense of temporal disjunction.

Our creative investigations have also been driven by postmemorial urges, by our need to find our own particular emotional truths related to our family histories. These urges are personal yet part of broader cultural and societal contexts. They are one form of the aforementioned juncture between memory and history; in this case not our own memory, but those of our parents' generation, refracted and augmented to serve our particular purpose: to better understand the mysterious past and where it belongs in our lives.

Creating our works has helped us towards attaining that understanding, as it has for others undertaking personally enriching media activity, whether as part of their professional practice or as amateurs (Stebbins, cited in Tarrant, 2003, p. 8). This is something the digital era has made much easier. Here we propose that, for us at least, the process of making (as opposed to the end result) has helped to mitigate the disruptive effects of family displacement. This has involved a change in focus from conserving and archiving, with all its issues of legacy, authority and power (Derrida, 1995) to embodied action.

To date our artefacts have included processed digital video works that explore the aesthetics of postmemory by focusing on loops and gaps. Between our two practices we have also undertaken installations, audio works, written texts and related online projects. A few examples of our video and installation work follow. The first is a screenshot from 'Displaced One,' a video art work by Berry using found family footage from 1940 with mobile phone footage shoot in 2007 (Figure 4).

Figure 4. Screenshot from 'Displaced One'

Next, in Figure 5, a screenshot from a video artwork by Berry, 'Unforgotten: Solovetsky Islands 1929-1933.' This project made use of a family photograph and an evolutionary algorithm that uses the photograph as a target, tracing the edge map of the photograph using fine lines to compose the image. The sequence of images produced by the algorithm was then assembled onto a timeline to create a video which showed the image materialising and dematerialising in a loop.

Figure 5. Screenshot from 'Unforgotten: Solovetsky Islands 1929-1933'

Figure 6 shows 'East and West' (2011), a three-screen video installation by Schutt juxtaposing snippets of digital videos, taken during visits to Germany and Poland. These videos feature books, an East German forest from the window of a moving train, and the Kolomac ruins. The work evokes movement to and from places of significance, drawing connections with changes in the former eastern and western borders of the German Reich.

Figure 6. 'East and West'

'Untitled' (2011) is an installation work by Schutt exploring stories of identity, displacement and exile. Mimicking a museum display case, but built using cardboard, packing foam and other impermanent materials, its selection of childhood, family and fieldwork artefacts are linked by unspoken connections and resonances.

Figure 7. 'Untitled'

THEORISING WHAT WE DO

So how to conceptualise this activity as research? We encountered two related problems here: the lack of delineation between our identities as researchers, creative media makers and subjects, and the eclectic nature of the work we do.

In regard to the former, anthropologist Clifford Geertz describes fieldwork thus: "Being There is a postcard experience. It is Being Here, a scholar among scholars that gets your anthropology read...published, reviewed, cited, taught" (1988, p. 130). But for us whose fieldwork has involved our own family history, the delineation between 'there' and 'here' is less than clear. Although it is impossible for us to 'be there' at a geographical place in the past, our close familial connections have turned what might otherwise be such a 'postcard experience' into something far closer to home.

In order to frame such work as 'research,' the researcher's challenge is, as Geertz (1988) states, in 'negotiating the passage from what one has been through 'out there' to what one says 'back here.' This he says, is a literary challenge, of 'rendering your account credible through rendering your person so ... To become a convincing "I-witness," one must, so it seems, first become a convincing "I"' (p. 79). And what makes a convincing research 'I'? Some might say a credible research persona involves an assumption of objectivity and distance from the subject, as encapsulated in the expression 'postcard experience.' For us however, this has not been the case. Indeed, Geertz's statement hints at the artifice involved in such an enterprise.

Instead, we perceive our persona as one who is open about overlaps in researcher/subject identities, pointing out that the 'I' (whether researcher, artist or subject) is fundamentally a 'theoretical manoeuvring' (Probyn, 1993, p. 83) connected with the idea of identity as a fluid, ever-emerging construct, or 'identity as becoming' (Hall, 1990). This kind of researcher persona embraces the foregrounding of subjectivity (and its associated and welcome acceptance of ambiguity) as proposed by qualitative research approaches such as autoethnography which are interested in '... people in the process of figuring out what to do, how to live, and the meaning of their struggles' (Bochner & Ellis, 2006, p. 111).

This persona would also attempt to ensure that the work of the 'I' is tied to broader social and cultural discourses in order to avoid the pitfalls of solipsism and narcissism in forms of research that elevate the 'I' to centre stage, as identified by qualitative researchers such as Riessman (2007) and Chang (2008).

In terms of the second problem – the eclectic nature of our work – rather than choosing one of many possible theoretical frameworks for our work, we have adopted the approach of 'travelling concepts' in which 'you do not conduct a method: you conduct a meeting between several, a meeting in which the object participates so that, together, object and methods can become a new, not firmly delineated, field' (Bal, 2008, p. 1).

Cultural theorist Griselda Pollock describes this as 'research as encounter,' stating that the 'theoretical turn' of the 1970s and 1980s has had its day, and that academic analysis is in need of a revamp through what is described as 'transdisciplinary encounters with and through concepts' (2008, p. xv). These concepts circulate 'between different intellectual or aesthetic cultures, inflecting them, finding common questions in distinctively articulated practices' (2008, p. i). Instead of being fixed, concepts are tools of intersubjectivity that facilitate discussion and debate (Bal, 2009), and 'may become attached quite rapidly to diverse phenomena including texts, practices and cultures' (Radcliff, 2008, p. 35).

This chapter has briefly touched on a range of travelling concepts: Lieux de Mémoire, life mapping, hauntology, postmemory, the aura and Being There. Of particular interest to us is the concept of hauntology. Originally coined by Jacques Derrida in *Spectres de Marx* (1993), the term refers to the spectre as a cipher for the unsettling of the present by unresolved, repressed or malevolent aspects of the past. The concept has since been used to explore a range of social, political and personal questions (Davis, 2005). Sociologist Avery Gordon (1997), for instance, has deployed the idea of 'haunting' to examine the effects of socio-political power and control, and to suggest avenues for resistance and renewal. We have also seen how the concept of hauntology can apply to our use of the (also well-travelled) term postmemory. Others have made related connections, such as Lisa Gye's (2003) use of hauntology to align genealogical research with ghost hunting, and Barbara Gabriel's work on memory, in which she invokes the notion of a haunted subject who yearns for a mythical lost homeland, quoting in the process Julie Kristeva's evocative questioning: 'How do we confront that which we have excluded in order to be, whether it is the return of the repressed or the return of the strangers?' (Kristeva, cited in Gabriel 2004, p. 149).

As a recent article in *The Guardian* newspaper points out, the idea of hauntology has since travelled widely, albeit in a mutated (and still mutating) form. These days it 'inspires many fields of investigation, from the visual arts to philosophy through electronic music, politics, fiction and literary criticism. At its most basic level, it ties in with the popularity of faux-vintage photography, abandoned spaces and TV series like Life on Mars' (Gallix, 2011). In its new life as a popular culture 'meme,' hauntology comes back to us, having picked up an intimate association with the creation of digital media (particularly those evoking a sense of a lost past) and the cultural effects of modernity, thereby resonating with another familiar travelling concept, Nora's Lieux de Mémoire:

> As a reflection of the zeitgeist, hauntology is, above all, the product of a time which is seriously "out of joint" ... There is a prevailing sense among hauntologists that culture has lost its momentum and that we are all stuck at the "end of history." Meanwhile, new technologies are dislocating more traditional notions of time and place ... nothing dies any more, everything "comes back on YouTube or as a box set retrospective" ... This is why "retromania" has reached fever pitch in recent years. (Gallix, 2011)

Again we see the point of rupture between memory and history, the sense of loss generating acts of creation that embody a yearning for a mythical past. And again, the presence of the spectre, never fully alive nor fully dead, now representing the problematic of the digital era itself, as much as the artefacts produced by it or the things those artifacts represent.

CONCLUSION

Here we would like to briefly offer two more examples of videos made by other people that cast further light (and corresponding shadows) on the discussion thus far. The Internet Archives, an online repository of public domain media, contains a digitised version of an intriguing 1938 home movie called *Vagabonds Abroad*. This silent amateur film was taken, using the available consumer technology of the time, by a young, seemingly well-to-do American couple on their grand tour of Europe and North Africa, where they took in both Munich, the city where Schutt was born, and Berlin, to where Berry's Russian family emigrated in the 1930s. Their on-the-fly, off-the-cuff but still highly framed visual account accidentally captured, like an insect in amber, a society and continent on the verge of irrevocable change. The couple wanted to create an informal narrative of their European adventure to remember and, we presume, to show their family and friends. But in doing so, they generated something unexpected, something that had the power to haunt lifetimes later, breaking the banks of its context in a way that nobody could have expected. In acting to turn their memory into their history, they have inadvertently created a spectre for their future.

The second example is, like ours, a research project that is also a family story and a creative endeavour. Patrick Tarrant's *Planet Usher* project (2003) 'tells the story of, and through, the home video archive of the artist's brother, Peter, a man who was born deaf, took 20 years of home videos, and has slowly gone blind due to the effects of Usher Syndrome' (Tarrant, 2005, p. 1). In engaging with 'the fantasy that new media might somehow revive the lost home video archive,' Tarrant states that '… it is productivity, not simply loss, that is at work here, and the figure for this productivity is the "remembered home movie" (2005, p. 1). This 'live' element is the process of creating the Tarrant family video archive, an undertaking of ongoing interest to both Tarrant and his brother because 'it continues to privilege an ethos of production over conservation, the latter proving itself to be a perilous pursuit and the former being an attempt to render the home in such a way that makes it available in the space of imagination for producers and audiences alike' (p. 4).

What these two examples show is that whereas history may aim to construct a definitive and ultimately atrophied narrative about the past, production as an embodied experience can be about generating new possibilities for remembering. This is why creative activity is so important to us. The European couple sought to capture the memories of their journey, but in doing so their 'I' inadvertently captured poignant snippets of a lost, living Germany (and Europe) that no other camera would have caught in its lens. It created a history unlike other histories, one of immense value because it was not conceived and framed as history. For Tarrant, his video project was 'a process of memory-making capable of generating an afterimage in a space beyond the time of images.' This 'afterimage' is as much an outcome as the artifact itself (p. 8), and points to the importance of the process, rather than the result.

In conclusion, in order to make sense of our lived experience of postmemory, we interrogated the discourses of memory as well as those of hauntology, history and ethnography. Again and again, these discourses took us to familiar places: the memory/history rupture, and the role of the digital in both widening and bridging this rupture. Meanwhile, our own (digitally-driven) creative research into our personal encounters with postmemory took us to places that already existed in our imaginations and had been haunting us in ways we had not always consciously understood. We made creative works so we could meet and talk with those ghosts, and thereby explore possible aesthetics of postmemory through the use of loops, gaps and absences. We found that this approach to research, which focused on the process of creation rather than any artefacts created, enabled us to uncover and recover our relationships to our own ghostly pasts. Paradoxically, we also ended up with creative artefacts that served to express our experiences of postmemory to ourselves and, hopefully, to others.

REFERENCES

Bal, M. (2002). *Travelling concepts in the humanities: A rough guide* (Green College Lecture Series). Toronto: University of Toronto Press.

Bal, M. (2009). Working with concepts. *European Journal of English Studies, 13*(1), 13-23.

Barthes, R. (1980). *Camera lucida: Reflections on photography* (2nd ed.). London: Hill and Wang.

Benjamin, W. (1932). A short history of photography, trans. by Phil Patton. *Artforum, 15*(6), 46-51.

Benjamin, W. (1932b). *Berlin childhood around 1900*. Cambridge, MA: Belknap Press.

Briggs, M.A. (2008). The return of the aura: Contemporary writers look back at the First World War photograph. In A. Kuhn, & K.E. McAllister (Eds.), *Locating memory: Photographic acts* (pp. 113-134). New York: Berghan Books.

Chang, H. (2008). *Autoethnography as method*. Walnut Creek, CA: Left Coast Press.

Chare, N. (2008). Puncture/punctum: Out of shot. In G. Pollock (Ed.), *Conceptual Odysseys: Passages to cultural analysis* (New Encounters: Arts, Cultures, Concepts), London: I.B. Tauris.

Davis, C. (2005). État présent: Hauntology, spectres and phantoms. *French Studies, LIX*(3), 373-379.

Denzin, N.K. (2006). Analytic autoethnography, or deja vu all over again. *Journal of Contemporary Ethnography, 35*(4), 419-428.

Derrida, J. (1993). *Spectres de Marx*. Paris: Galilee.

Derrida, J. (1995). *Archive fever*. Chicago: University of Chicago Press.

Ellis, C.S., & Bochner, A. (2006). Communication as autoethnography. In G. Shepherd, J. St. John, & T. Striphas (Eds.), *Communication as ...: Stances on theory* (pp. 110-122). Thousand Oaks, CA: Sage

Gabriel, B. (2004). The unbearable strangeness of being; Edgar Reitz's heimat and the ethics of the unheimlich. In B. Gabriel, & S. Ilcan (Eds.), *Postmodernism and the ethical subject*. Montreal/ Kingston: McGill-Queen's University Press.

Gallix, A. (2011). Hauntology: A not-so-new critical manifestation. *The Guardian Newspaper*. Retrieved from http://www.guardian.co.uk/books/booksblog/2011/jun/17/hauntology-critical.

Geertz, C. (1988). *Works and lives: The anthropologist as author*. Stanford, CA: Stanford University Press.

Gordon, A. (1997). *Ghostly matters: Haunting and the sociological imagination*. Minneapolis, MN: University of Minnesota Press.

Gye, L. (2003). Halflives, A mystory: Writing hypertext to learn. *Fibreculture Journal, 2*, New media, new worlds? Retrieved from: http://two.fibreculturejournal.org/halflives-a-mystory-writing-hypertext-to-learn.

Hall, S. (1990). Cultural identity and diaspora. In J. Rutherford (Ed.), *Identity: Community, culture, difference*. London: Lawrence & Wishart.

Hirsch, M. (2001). Surviving images: Holocaust photographs and the work of postmemory. *Yale Journal of Criticism, 14*(1), 5-37.

Hirsch. M. & Spitzer, M. (2006). What's wrong with this picture? Archival photographs in contemporary narratives. *Journal of Modern Jewish Studies, 5*(2), 229-252. Retrieved from: http://www.columbia.edu/~mh2349/papers/CMJS_A_174124.pdf.

Humphreys, M. (2005). Getting personal: Reflexivity and autoethnographic vignettes. *Qualitative Inquiry, 11*(6), 840-860.

Nora, P. (1989). Between memory and history: Les lieux de mémoire. *Representations, 26*, 7-24.

Pollock, G. (Ed). (2008). *Conceptual Odysseys: Passages to cultural analysis* (New Encounters: Arts, Cultures, Concepts). London: I.B. Tauris.

Probyn, E. (1993). *Sexing the self: Gendered positions in cultural studies*. London: Routledge.

Schutt, S., & Berry, M. (2011). The haunted photograph: Context, framing and the digitalised family story. *Current Narratives, 3*, 35-53.

Radstone, S. (2008). Memory studies: For and against. *Memory Studies, 1*(31), 32-40.

Riessman, C.K. (2007). *Narrative methods for the human sciences*. Thousand Oaks, CA Sage.

Tarrant, P. (2003). Planet Usher. Retrieved from: http://patricktarrant.com/sequences/index.htm.

Tarrant, P. (2005). *Home made: New media in the home mode. Speculation and innovation: Applying practice led research in the creative industries*. Queensland University of Technology. Retrieved from: http://www.patricktarrant.com%2Fmedia%2F%2FTarrantP.pdf.

Vagabonds Abroad (1938). [Video File]. Retrieved from: http://www.archive.org/details/EuropeanTrip.

Stefan Schutt
Work-based Education Research Centre
Victoria University, Australia

Marsha Berry
School of Media and Communication
RMIT University, Australia

LINDA SHALLCROSS. SHERYL RAMSAY AND
MICHELLE BARKER

QUALITATIVE INQUIRY AS TRANSFORMATION AND AGENCY

The black sheep and workplace mobbing

INTRODUCTION

This chapter discusses an action research project into the lived experience of the workplace mobbing phenomenon. The action research methodology is based on the exemplarian model (Coenen & Khonraad, 2003) from the Netherlands Group. This model requires positive outcomes for those immersed in the problem to reduce the adversity of their circumstances. The findings challenge the psychological perspective of the existing bullying literature that tends to focus on individual behaviour. This research, undertaken over a three year period with 212 participants, identified the dysfunctional nature of public sector bureaucracies and the power gained through gossip and rumour as some of the key emergent themes to explain the workplace mobbing problem. In addition, resistance, conscientisation, and agency were identified as the key to transformation for those targeted. The discussion focuses on the crystallisation phase of the exemplarian model where the participants identified themselves as the Black Sheep and adopted the motto that "a black sheep is a biting beast" (Bastard, 1565 or 6-1618, p. 90), reflecting a sense of empowerment, individual agency, and a sense of humour in dealing with the serious yet seemingly absurd reality of their situations. The identity of the Black Sheep was consolidated when the group organised a 2 day conference with over 200 attendees to discuss how best to prevent workplace mobbing. This self-affirming action was a proactive step towards metaphorically "biting back" at the problem. A number of positive outcomes were achieved including the conference with over 200 attending leading to national media coverage across Australia and additional interviews with magazines, newspapers, and radio.

METHODOLOGY

The methodology, embedded in action research, aims for transformation and the achievement of positive outcomes for all of those involved. The discussion is based on a three year study with 212 public servants from states and territories across Australia. The research was driven by those who self identified their workplace situation as one where mobbing had occurred. They found out about the problem

M. Vicars et al. (eds.), Discourse, Power, and Resistance Down Under, 121–133.

from media reports, based on earlier Australian research that explained mobbing as a five phased process to squeeze people out of their jobs. The phenomenon includes social isolation and expulsion from employment that most often results in long-term harm including psychological damage, loss of income, and long term unemployment. However, the methodological underpinnings of the study that highlight the role of qualitative inquiry, in this case exemplarian action research, in achieving transformational change, is first discussed.

The exemplarian model shares some common principles with participatory action research (PAR) including creating spaces where those who tend to be marginalised can become involved in change and resistance (Kemmis & McTaggart, 2005). Each of these methodologies can be described as interventionist, action-oriented, interpretive and concerned with developing purposeful knowledge through practical engagement with the world (2005). The exemplarian model, however, is explicit in regards to the role of the researcher, not only as a participant but as an equal participant immersed in the research problem to the same extent as the other participants (Boog, 2003, p. 426). While both methodologies focus on outcomes, the exemplarian model requires the achievement of transformative outcomes and the identification of exemplars that can be used by others in similar situations. This model may assist other researchers, similarly immersed in a problem, with an example of an emancipatory methodology that requires individual and collective agency to be exercised in the achievement of outcomes.

JOINT PARTICIPATION

The "jointness" (p. 439) of the relationship between "the researcher" and the other participants is indicated in the following email, where the researcher seeks advice from the Black Sheep about a suitable advocate to assist in a claim for worker's compensation in the Industrial Relations Commission (IRC).

> I had the expected rejection of my claim ... yesterday. The usual whitewash and claiming my injuries are due to 'reasonable management action' ... I can now appeal to the Industrial Relations Commission and then off to the Industrial Magistrate's Court.
>
> I've expected all the way along with this second claim that it has just gone through the motions as this really has to go to court.
>
> I'll be trying Legal Aid for support ... and if all that fails I'll be going to [the solicitor]. I've got 28 days to appeal so wish me luck. Dianne, do you think your advocate would be suitable for my trek through the Commission?

This email example highlights the open communication, reflexivity, and planning between the participants and 'the researcher' as a joint participant. This type of relationship gives voice to experience and promotes a sense of power, involvement and agency with others. Agency, in this context, refers to the human capacity to "act otherwise" and the continuous reflexive monitoring of human action (Giddens, 1987, p. 216). This level of collaboration required trusting relationships to be established and the development of action plans to motivate one another to achieve goals. Ideas were discussed, actions were trialled, and reactions were observed and learned from. The daily, and sometimes hourly, intensive reflections took place in a virtual space created with the assistance of the National Women's Justice Centre (NWJC) on a computer server. This enabled participants to confidentially reflect and record their thoughts, emotions, ideas, and stories, at any time of the day and night, throughout the three year inquiry. The methodology was empowering, not only for the Black Sheep, but also for the researcher, in taking risks with a number of unexpected and successful outcomes.

EXPLICIT KNOWLEDGE

The exemplarian action research methodology does not claim to be value free and requires the researcher to be explicit in acknowledging their own experience in the research context (Coenen, 2003). In this study, the researcher held a managerial position in the public sector where the covert bullying behaviours of a group of four or five staff came to her attention. In particular, the staff appeared to be harshly critical of any new member of staff. At some point, this group directed their hostilities, albeit covert, towards her, eg refusing to speak to her or to interact in a reasonable manner. Other behaviours included taking unplanned leave together, on those days when major activities had already been organised, leaving the service short staffed and requiring the cancellation of events. Offensive emails including one that, when opened, buzzed loudly, and flashed onto the computer screen repeatedly with the words appearing in large text, MY BOSS IS AN ASSHOLE.

The conflict escalated when, during the manager's absence, the group complained that she was a bully, resulting in her immediate removal from the workplace and subsequent investigation. While it was found, two years later, that she had no case to answer management had already sided with the group. Additionally, they had promoted the main ring leader to her position. Furthermore, another member of the group was successful in gaining the support of a government minister who raised the allegations in parliamentary question time, without notice, and under parliamentary privilege as follows:

> … has the complaint of [name] concerning the manager for Women's Infolink and the considerable supporting evidence … been investigated and has any action been taken against the manager and … are there plans to remove the root cause of those problems – the manager of Women's Infolink as well? I seek the leave of the House to table the supporting evidence for the Minister's information.

During the course of the investigation, it was revealed that a number of previous managers had also been targeted with similar behaviours. Indeed, the workplace morale had deteriorated to the extent that staff were openly hostile towards one another before the latest manager was appointed. This had been sufficiently serious for the department to engage a psychologist to work with the group on a regular basis to discuss and monitor appropriate workplace behaviours. The situation deteriorated upon structural relocation of the agency after a government election, and with associated loss of corporate memory. New departmental officers, unaware of the historical context, inadvertently, at least to begin with, joined in with the mobbing behaviours.

Some 12 months later, a paper on workplace mobbing was presented at a small, local conference. This presentation struck a chord with a journalist and a brief report was published in newspapers across Australia. The authors were contacted by many people who identified their experience as workplace mobbing. Additionally, they were contacted by national media and interviewed live-to-air. This exposure was followed with a large influx of emails from many others making similar claims. After hearing or reading the media reports, they said that they were able to name and identify what had happened to them. This gave impetus for further research driven by the demand for recognition and accountability for the harm caused by workplace mobbing.

INSPIRATION

Of particular interest to the Black Sheep was the mobbing experience of a former Chief Magistrate of Queensland, Ms Di Fingleton, who was sentenced to 12 months' imprisonment "with no recommendation for parole" on a charge of "retaliating against a witness" (Hunter, 2004, p. 145). The witness was a fellow magistrate whom the Chief Magistrate had attempted to discipline. The circumstances of the situation were published regularly in the media where she was identified as a "bully" and publicly humiliated for at least two years. She was later found by the Australian High Court, to have been wrongly imprisoned and was subsequently released after serving six months of her sentence. Another prominent woman, Dr Jocelynne Scutt, the then Commissioner of the Tasmanian Anti-Discrimination Commission, had also experienced mobbing in her capacity as Commissioner. In her final 2004 Annual Report to the Tasmanian Government, Dr Scutt reported that her employment experience at the Commission was one of workplace mobbing, where she had been subjected to:

> bullying, pressure and other means … to curtail the integrity of the Commission's decision making" in regard to dealing with complaints. She cited her experience as including incidents of "defamation on several occasions … false imprisonment, abuse of process, contempt of court … and victimization" to the extent that she required police protection. (Scutt, 2004, p. 9)

These two examples, involving well-regarded women were perceived by the Black Sheep to be inexplicable travesties of justice with which they had some empathy. These two high profile cases provided continual reminders, if any were needed, that this research problem was worth investigating in an effort to address future similar miscarriages of justice.

EMERGENT PROPOSITIONS

Overall, the findings indicate that the organisational response to mobbing is one that condones and perpetuates the problem by blaming the target and excusing management behaviours, regardless of the damage caused to individuals, their families, and to the wider community. Some of the behaviours include lengthy drawn-out investigation processes, continual psychiatric assessments, the rejection of claims for compensation and compulsory retirements on the basis of psychological ill health. The outcomes achieved by the participants in this study, despite the ongoing pressure to acquiesce, were therefore significant in reducing the level of damage caused.

Table 1. Emergent propositions

Propositions	**Themes**
Public sector culture is dysfunctional whereby employment survival requires conformity, submission, and silence. (Organisational culture) Workplace gossip, rumour, hearsay and innuendo are influential forms of power in public sector organisations. (Gossip and power) While guidelines, detailing principles of natural justice and due process have been developed, these are not enforceable and do not match with public sector practice. (Organisational justice) Support systems for targeted workers are not neutral and tend to act on behalf of the employer to the detriment of the employee. (Support systems)	Theme 1: Expulsion
There may be a gendered pattern to workplace mobbing where women are not only more likely to be targeted but are also more likely to perpetrate acts of mobbing towards other women. (Gender and mobbing) A contributing risk factor for being targeted appears to relate to belonging to a cultural minority, that is, being an outsider or different to the dominant culture. (Exclusion)	Theme 2: Exclusion
Workplace mobbing is a distinct form of workplace violence, and to give voice to those targeted, the phenomenon first needs to be recognised and understood. (Naming the problem) The absence of specific legislation to address the phenomenon appears to maintain a system that denies legal remedies to those targeted (Legislation) Survival of mobbing most likely requires those targeted to take risks in the pursuit of options to reduce the adverse impact of the problem. (Transformation)	Theme 3: Transformation

The findings are expressed as nine emergent propositions grounded in the voice of the participants. These include organisational culture, gossip and power, organisational justice, support systems, gender and mobbing, exclusion, naming the problem, legislation, and resistance as listed in Table 1.

AGENCY

The outcomes surpassed the participants' expectations with several achieving financial settlements, other achieving higher education goals, the organisation of a successful conference, and changed organisational practices in some instances. These were achieved through individual acts of agency including lodging claims for compensation, grievances and seeking legal remedies. At the community level, outcomes included an increased awareness of the problem as measured by the number of visits to the *workplace* mobbing website and increasing requests for information and assistance. This was achieved through the distribution of information through the electronic and print media and debate, for example, in the Western Australian Parliament ("*Royal commissions (powers) Amendment Bill 2004*") calling for organisational change. Some other outcomes achieved at the departmental level included revised guidelines for compulsory ill health retirement, that is Section 85 of the Queensland Public Service Act, and revised guidelines for carrying out investigations, as well as the inclusion of some mobbing behaviours in workplace harassment guidelines, eg including spreading of rumours and gossip.

The individual acts of agency included lodging grievances, pursuing workers compensation claims, and seeking assistance from medical, health, and legal professionals. The media was also contacted on occasions to progress the achievement of outcomes. To alleviate the severity of their financial circumstances, the participants attempted to access support systems for injured workers including workers compensation, rehabilitation programs, and social welfare benefits. However, the participants found that they were unlikely to receive assistance, for example, of the 17 participants claiming workers compensation only three were successful. The experience with support systems led participants to conclude that they had been naive in their assumptions that they could rely on commonly accepted standards of ethical and fair behaviour. For example, although they would have preferred conciliatory methods of resolution, these options were not made available to them. Rather, they were subjected to adversarial processes, including suspensions and investigations that escalated conflict. Consequently, the participants realised that if they were to survive financially, they would need to engage in the adversarial methods of the systems world. These actions, described as deontic retaliation, included the exercise of agency in lodging grievances, claiming workers compensation, accessing superannuation entitlements, taking extended leave, and pursuing legal options. With advocacy support from the Black Sheep, individual acts of agency became possible whereby participants were able to reduce the adverse consequences of the problem.

One of the propositions is that support systems are not neutral and tend to act on behalf of the employer to the detriment of the employee. This theory reflects the experience of the participants in seeking support from the medical and health professions, the legal system, and the industrial relations system. The participants discovered that espoused ideals of justice, fairness, and equity were not reflected in practice and that access to entitlements was dependent on winning legal arguments through the industrial court systems. In order to survive the post expulsion financial consequences, the participants realised that they would need to re-engage with the systems world to access compensation, superannuation, and/or rehabilitation programs.

The lengthy process of claims and appeals experienced by the participants extended for months and years, for example, one participant while initially making a successful claim was subsequently involved in five years of reviews at the end of which her claim was rejected. However, while claimants appear to have little difficulty in proving long-term psychological damage, rather the issue becomes one of proving through the court systems, that management actions are unreasonable in order to access rehabilitation or to obtain compensation. Consequently, while many claims are made, only a few are accepted as indicated by the rulings listed on commission websites, where claims tend to be rejected on the basis of reasonable management action (see for example, the recent decisions of the Queensland Industrial Relations Commission at www.qirc.qld.gov.au).

HEALTH OUTCOMES

While the emphasis in this discussion is the achievement of emancipatory outcomes for economic survival, another outcome identified is the health benefits gained from exercising agency. Typical comments made by participants include:

> I am feeling more in control, have more good days than bad … since making contact with [black sheep] I have to say my health has made a marked improvement.

However, before being informed fully about workplace mobbing, this participant reported that she:

> had to go back to the Dr this morning, the sleeping drugs simply are not working, they have now put me on a drug called mirtazapine … I look dreadful, I carry excess baggage under my eyes, there has to be some end to this nightmare.

However, through a process of conscientisation, she commented 18 days later on 30 July that:

> [Your information] has done more for me than anything else. I guess just knowing I was not being a drama queen, this sort of thing has happened to many others, and it has been acknowledged by someone in the know [*sic*]. I am now in the phase of just trying to get on with my life.

The following day, this participant reported to the workers compensation agency that, "Out of all the help and assistance I have received from anyone, this information about [workplace mobbing] has been the most worthwhile." Subsequently, the agency replied that they understood there had been a "good therapeutic outcome" and on that basis agreed to fund an assessment of her workplace mobbing experience.

FINANCIAL OUTCOMES

Some of the participants obtained financial settlements. These included voluntary early retirement (VER) packages although these were not generally available to all staff at the time. VERs, in these circumstances, amounted to the equivalent of the participant's salary for one year. Other financial settlements negotiated through the industrial commissions were equivalent to approximately six weeks salary. Sometimes settlements were a combination of a VER and a comparatively small payment on the basis that other legal proceedings be discontinued. For example, one participant was offered AUD 11,000 to discontinue her complaint; however, because of a requirement in the deed of settlement that she forego any other legal action and not make public her situation, she refused to sign.

COMMUNITY AWARENESS RAISING OUTCOMES

During the 3 year study, members of the Black Sheep initiated and/or otherwise contributed to reporting on the workplace mobbing problem in various media forms across Australia. This including organising a conference for others who may have experienced the problem, publishing stories on line, assisting others, and increasing community awareness through a number of media reports as indicated in Table 2.

REVISED GUIDELINES FOR COMPULSORY PSYCHIATRIC ASSESSMENT

During the course of this research, participants became increasingly aware that to survive the experience of workplace mobbing they were compelled to exercise their individual agency. The duration and intensity of assaults was a common theme and the experience of one participant, typifies those of many others and is used here as an exemplar to highlight the process of transformation and agency.

Three attempts were made to compulsorily retire this participant on the basis of psychological ill health. She was directed to appointments with psychiatrists for this purpose in two consecutive years. A chronological analysis of her experience indicates that during Year 1, while she was subjected to 17 psychological assaults she responded with acts of individual agency only five times. However, after collaboration with Black Sheep during Year 2, her rates of response increased to 11 until she was able to finally negotiate a VER package in Year 3. An organisational outcome achieved in this process included the introduction of revised guidelines for referrals of psychiatric assessment under Section 85 of the Public Service Act 1996 (Qld).

Table 2. Media reports explaining workplace mobbing

Year 1	Date
Sunday Mail (Queensland) – Pssst! Heard the latest about vicious gossip	19-Oct
Sunday Mail (Adelaide) – Mobbing can kill a career	19-Oct
Northern Territory News – Gossip can be lethal weapon	19-Oct
Sunday Tasmanian (Hobart) – Office gossip (mob) can kill a career	30-Oct
Life Matters ABC Radio Interview with Kim – Work with Rebecca Gorman	1-Nov
Gold Coast Bulletin – Work's hidden danger	1-Nov
Brisbane Courier Mail – Mob rule at work	17-Nov
Year 2	
Brisbane Courier Mail – Wired for sound and fury	7-Feb
Gold Coast Bulletin – Severe bullying may cause serious illness	24-Apr
Brisbane Courier Mail – Help for those done over by the mob	28-Sep
Life Matters ABC Radio – Interviews re Mobbing Conference	14-Oct
Radio 2GB Sydney – Interview with Alan Jones	15-Oct
Sunday Mail Brisbane – Psych tests dished out as punishment	31-Oct
Sunday Mail Queensland – Storm grows over psychiatric tests – 'hitmen'	14-Nov
Sunday Mail – Call for ban on psych testing	21-Nov
Globe and Mail Canada – Mobbing: bullying's ugly cousin	12-Dec
Sunday Mail Queensland – Public service bosses hear darkest secret	8-Dec
Year 3	
Australian Health & Safety Matters Magazine – Workplace mobbing: An OHS issue (Feature article)	1-Jan
Brisbane Courier Mail – Minister's staff member wins settlement over sacking	14-Jan
Brisbane Courier Mail – I'm a victim of mob justice	28-Jun
Financial Review – When the office rumour mill gets ugly	19-Jul
Townsville Bulletin – Problems at the top	11-Aug
Brisbane Courier Mail – Harassment culture starts at the top	28-Sept
Brisbane Sunday Mail – Government accused of playing unfair mind games	11-Aug

This exemplar demonstrates that individuals can affect some degree of organisational change. While the guidelines were reportedly welcomed by some psychiatrists as a "step in the right direction," nevertheless they would have preferred that these "be abolished altogether" or at least "rewritten [so that] referral for assessment [can] only occur with a person's consent" (Passmore, 2004b). The intention of the revised guidelines, it was reported, was "to stop government agencies misusing legislation to get rid of whistleblowers and workers on dubious psychiatric grounds" (Passmore, 2004c).

Another emergent theme was in reference to the use of "hired guns" on a regular basis to conduct compulsory psychiatric assessments that tended to favour the employer. The psychiatrist commented in a newspaper report that some of his colleagues had saved insurance companies "lots and lots of money" as a result (Passmore, 2004c). The Black Sheep's courage in voicing the issue with a newspaper reporter resulted in some of these issues being brought to public attention and contributed to the generation of additional media reports as listed in Table 3.

Table 3. Media reports lobbying against compulsory psychiatric retirement

Date	Incident
17 Jan Yr 1	Mental health services abused by managers – Radio interview 7.30 Report
30 Oct Yr 1	Workplace mobbing radio interview – Life Matters, Radio National, ABC
14 Oct Yr 2	Paper presented at workplace mobbing conference – Susan Moriarty
31 Oct Yr 2	Media: Psych tests dished out as punishment
2 Nov Yr 2	Psych testing used to sack whistleblowers – Radio interview 612 ABC
7 Nov Yr 2	Psych testing hotline: Support group to aid bullied public servants
14 Nov Yr 2	Storm grows over psychiatric tests – 'hitmen'
21 Nov Yr 2	Call for ban on psych testing
23 Nov Yr 2	Hansard Queensland parliament: Public servants psychiatric testing
8 Dec Yr 2	Public service bosses hear darkest secrets
7 Nov Yr 3	Guidelines set after referrals queried
7 Nov Yr 3	Government accused of playing unfair mind games

One of the issues raised by the Black Sheep was in relation to the requirement by some psychiatrists to complete a questionnaire asking about:

> sexual preference, whether you are adopted, criminal record, all sorts of drug use. the quality of all the relationships in your life going back to school years, reasons you left all your previous jobs, your complete medical history, psychiatric history, whether there is a history of psychiatric illness in your family, whether you or your parents are divorced and so on. (Flegg, 2005)

The matter was also raised with the leader of the Opposition who subsequently raised the matter in parliamentary debate as follows:

> I want to speak about Section 85 of the Public Service Act and how it empowers a public sector employer to order an employee to undergo a physical or mental examination with a doctor and also provides that the employer under the Act may receive a copy of that report.
>
> If a superior feels threatened by a talented, performing work colleague or if there is a personality conflict or a superior's human resources skills are inadequate to sort out the staff, Section 85 can be used as a handy tool (Flegg, 2004, p. 3606)

Furthermore, the damaging process of referral was described as one where those targeted:

> are not given appropriate avenues through which to respond. They are not permitted to be accompanied at the examination or to record the proceedings. They have limited rights of appeal. They have to live with the knowledge that this information can come into the possession of those with whom they must continue to work. Then on top of all of this, they can be referred on multiple occasions despite a long and satisfactory career in the Public Service and the clear absence of psychiatric illness. (Passmore, 2004a)

The issue gained momentum, generating further community interest and a support group was established. Additionally, criticisms of the practice appeared in the newspaper editorial column. Following up with the Member of Parliament, three months later, one of the action research participants, forwarded a six page list of questions that had been provided to her at the commencement of a compulsory psychiatric assessment. The matter was raised in the parliament shortly thereafter as follows:

> We see the misuse of Section 85 of the Public Service Act to enable departmental officers to compulsorily force public servants to undergo psychiatric testing, compelled to have a degrading and highly personal six-page psychiatric questionnaire and allowing reports of the most intimate aspects of people's mental health to be reported back to other members of the Public Service. (Flegg, 2005)
>
> The Premier confirmed that 243 employees have been referred for medical assessments under Section 85 – a full half of these for supposed mental health issues. The Premier naturally does not go on to say that almost none of the 120 or so public servants put through this degrading, intimidating and bullying process suffer any mental illness. Nor does he explain why use of this section has skyrocketed by hundreds of percent each year over the last few years. (Flegg, 2005)

The public criticism generated through the media and the parliament about compulsory psychiatric assessments resulted in the introduction of new guidelines to improve the process.

SUMMARY CONCLUSION

This chapter explores the application of an innovative action research methodology used to explore the problem of workplace mobbing. The methodological underpinnings of the study highlight the role of qualitative inquiry in achieving transformational change for participants through conscientisation and the exercise of agency. The exemplarian model highlights agency, resistance, and the need to take risks in reducing the adverse impact of problematic workplace behaviours.

REFERENCES

Bastard, T. (1565 or 6-1618). *Chrestoleros; seuen bookes of epigrames.* New York: Burt Franklin [1967].

Boog, B. (2003). The emancipatory character of action research, its history and the present state of the art. *Journal of Community & Applied Social Psychology, 13*(6), 426-438.

Coenen, H. (2003). Theoretical remarks to the model of exemplarian action research. In B. Boog, H. Coenen, & L. Keune (Eds.), *Action research: Empowerment and reflection.*

Flegg, B. (2004). *Public servants: Psychiatric testing* [MP Mogill] (23 November ed.). Brisbane: Hansard Publications.

Flegg, B. (2005). Appropriation (Parliament) Bill (No. 2) (51st Parliament ed.). Brisbane: Hansard, Queensland Government.

Freire, P. (1993). *Pedagogy of the oppressed.* New York: Continuum Books.

Giddens, A. (1987). *Social theory and modern sociology.* Oxford: Polity Press.

Giddens, A. (1999). Elements of the theory of structuration. In A. Elliott (Ed.), *Contemporary social theory.* Malden, MA: Blackwell Publishers.

Hunter, R. (2004). Fear and loathing in the sunshine state. *Australian Feminist Studies, 19*(44), 145-157.

Kemmis, S., & McTaggart, R. (2005). Participatory action research: Communicative action and the public sphere. In N. Denzin, & Y. Lincoln (Eds.), *The Sage handbook of quality research* (3rd ed.). Thousand Oaks: Sage.

Passmore, D. (2003, October 18). Psst! Heard the latest about vicious gossip? *The Sunday Mail,* p. 38.

Passmore, D. (2004a, November 7). Psych testing hotline: Support group to aid bullied public servants. *Courier Mail,* p. 42.

Passmore, D. (2004b, October 24). Psych tests dished out as punishment: Claims of office bullying. *The Sunday Mail,* p. 18.

Passmore, D. (2004c, October 31). Storm grows over psychiatric tests used against public servants – Doctor tells of medical 'hitmen.' *The Sunday Mail,* p. 40.

Royal commissions (powers) Amendment Bill 2004. (2004). (25 August ed.). Parliament of Western Australia Hansard Search Index

Scutt, J. (2004). *Celebrating difference, embracing equality: Fifth annual report 2003-2004.* Hobart: Department of Justice, Tasmanian Government.

Linda Shallcross
Griffith University, Australia

Sheryl Ramsay
Griffith University, Australia

Michelle Barker
Griffith University, Australia

ERICA SOUTHGATE

ETHNOGRAPHY AS THE (OCCASIONAL) ACT OF REFUSING CORRECT KNOWLEDGE AND SECURE UNDERSTANDING

Three tales from three fields

INTRODUCTION

Feminism has long concerned itself with the conditions under which knowledge is produced. Poststructural feminism has focused on the problem of working within and against the foundational categories of knowledge such as voice, authenticity and agency. This chapter examines the benefits of an empirical approach to working with and against foundational categories by spotlighting particular occasions when the production of knowledge from ethnography becomes difficult because of intersubjective, emotional and ethical reasons. The chapter uses field notes to examine three epistemological dynamics emerging from three different research projects. These dynamics are knowing-but-not-telling, not-knowing, and not-speaking-of-knowing. The implications for poststructural feminism in adopting an empirical approach to knowledge production are discussed.

A HAPPY POSITIVIST COMES CLEAN

This chapter is about my experience as an ethnographer conducting research in three different research fields over a decade. Each research project focused on "finding" evidence to address a social problem. Empirically, this chapter draws on field notes from studies in: (1) public health research on risk practices for blood borne virus transmission for people who inject drugs (PWID); (2) public health research on the health needs of sex workers; (3) educational research on community engagement in low socioeconomic status (SES) school communities. Pseudonyms are used for people, places and organisations.

Arguably, each of these fields is imbued with powerful positivist understandings regarding the "problem" to be investigated and the role of research and the researcher in finding evidence to help "solve" the problem (Israel, Schulz, Parker, & Becker, 1998; Lather, 2007). This positivism manifests itself in a familiar set of actions: The researcher sets off to collect reliable, verifiable, generalisable evidence to help fix a social problem identified by policy makers who fund the research (Bacchi, 2010). Once "found," this evidence is written up into a "realist tale" (Van Maanen, 1988) with recommendations on what to do about the problem. While it has been possible to write alternative research tales, these are less

M. Vicars et al. (eds.), Discourse, Power, and Resistance Down Under, 135–149.

common and harder to write, particularly given the resurgence of "scientism" which promotes positivist understandings about what good, hard evidence is and the imperative for good researchers to find it (see Clegg, 2010; Eisenhart & Towne, 2003; Hammersley, 2007, 2008; Lather, 2006, 2007; Trotter, 2003).

One of the most concerted challenges to positivism has come from feminists, including those of the poststructuralist ilk (for an overview see Nagy Hesse-Biber, 2011). Lather (2007) voices some of the key questions raised by poststructural feminism when she asks:

> Given the indignity of being studied, the violence of objectification ..., what are the ethics of such pursuits? What work do we want to inquire into? To what extent does method privilege findings? What is the place of procedures in the claim to validity? What does it mean to recognise the limits of exactitude and certainty, but still have respect for the empirical world and its relations to how we formulate knowledge? What is left for science in the era of blurred genres? (p. 39)

Following this logic, Lather (2007, 2009) and others like her (Koro-Ljungberg, 2010; Mazzei, 2007; White & Drew, 2011) suggest that feminist research needs to work within and against the ruins of its foundational categories such as voice, identity, agency, authenticity, empathy and experience (Koro-Ljungberg, 2010; Mazzei, 2007). In this chapter I respond to the challenge to work within and against the ruins by interrogating moments when my positivist desire to find good hard evidence dissolved into a mode of thought and action that was "emotive, inexact, dispersed and deferred" (Lather, 2007 p.23). These moments represent a *refusal* to seek out the coveted "lost objects" of scientific inquiry – the ideas of correct knowledge and security of understanding (Lather citing Britzman, 2007, p. 18). The epistemological reasons for examining occasions of refusal sit well with Foucault's (2003) idea of critique:

> A critique does not consist in saying that things aren't good the way they are. It consists in seeing on just what type of assumptions, of familiar notions, of established and unexamined ways of thinking the accepted practices are based ... To do criticism is to make harder those acts which are now too easy. (p.172)

I provide a critique of certain occasions in ethnography where correct knowledge and secure understanding were not accomplished. Each tale from the field documents a particular dynamic of refusal. These dynamics allow for the scrutiny of unexamined ways of thinking about research practice as the seamless (positivist) uncovering or (social constructionist) production of useful evidence. While reflexive ethnographic accounts have become more common, Blackman (2007) and others (Punch, in press; Suri, 2008; Woodthorpe, 2009) point out that there is still a reluctance on the part of researchers to disclose how knowledge production is influenced by the emotional encounters, hidden struggles and challenges of the fields in which they ply their trade. Blackman (2007) calls for the writing of hidden ethnography or those stories that are part of the oral culture of researcher but are rarely published for fear of leaving the researcher professionally vulnerable to charges of collusion, advocacy and unethical behaviour. The call for hidden ethnography resonates with the Foucauldian endeavour to uncover epistemological assumptions and reject familiar ways of thinking and knowing.

While autoethnographic in style (Reed-Danahay, 1997), this chapter focuses on key research moments where political, ethical and epistemological concerns intersect to produce a dynamics of refusal. Lather (2007, p.9) suggests that ethnography is well placed to learn from the "instructive complications" of knowledge production. To come clean, much of the research I produce does not examine the dynamics of refusal or expose the "instructive complications" of producing knowledge. Working within the areas of public health and education, both permeated with positivist ideas about what constitutes legitimate research, I gladly beaver away using foundational categories and writing up realist tales for participants, practitioners and policy makers, alike. To borrow a phrase from Foucault (1972, p. 125), I am a perfectly happy positivist, remaining reasonably confident in the empirical "science" of my research (Southgate et al., 2003; Southgate & Hopwood, 2001; Southgate, Weatherall, Day, & Dolan, 2005), while sometimes critiquing it through reflexive writing (Southgate, 2003a, 2003b, 2011). I am a poststructural feminist who is not driven by a desire to constantly refuse the idea that research can produce correct knowledge and security of understanding. Instead of being in a constant ontological and epistemological state of refusal, I prefer to examine the few occasions, over a fairly long research career, when a dynamics of refusal emerged from the field, in all its empirical (material) senses, to provoke lingering and unsettling epistemic tensions. This chapter is about those occasions of refusals and their instructive complications.

FIELD 1: SANDAY PUBLIC HOUSING ESTATE

Field diary extract – September 2001[i]

> I just got back from doing surveys at Sanday Housing Estate and I'm still a bit hyped. I went with Vera to do some surveys for a study on Hepatitis C transmission among PWID. (Hepatitis C is transmitted through infected blood by sharing needles and other injecting paraphernalia like tourniquets, spoons, swabs, water ampoules and possibly through environmental contamination). Vera works for a grass roots health organisation which has many users as clients. She sometimes partners with researchers to help out with recruitment for research. Vera is an ex-user and so knows how to tap into this hidden population. Vera made contact, through a friend of a friend, with Chanel who lives in Sanday, a large housing estate on the outer rim of a city. Chanel is letting us do surveys at her house and helping to round up people to do the survey.
>
> Chanel shows us into a neat living room that has a newish dining table and lounge suite in it. She tells us we can do the surveys at the dining table. I set

> myself up at one end of the table and listen and nod while Vera and Chanel chat. Five people arrive in the first half hour and take a seat, waiting room style, on Chanel's lounge while Vera and I administer the survey to each individual. A couple of them pop upstairs to use the toilet. About every ten minutes there's a knock at the door and a few more folk wander in to wait their turn. Some go upstairs. After about an hour and half, there's a knock at the door and a young bloke flies in and says, "Robbo's on the warpath. He's running around with a bit of four-be-two (timber) looking for the cunts who ripped him off." This news disrupts the flow of proceedings as the group discusses what to do if Robbo shows up. "Don't worry," Chanel says, "If he comes here I'll rip his guts out." Two people decide to leave and Chanel goes out into the back yard and looks over the fence to see if Robbo is roaming around. She says, "I didn't see him but I think he's out there."
>
> By now it's been a couple of hours since we started work and I ask Chanel if I can use her toilet. Glancing around at the lounge room she says to the two people sitting in the lounge, "Is it OK?" They nod and she says, "Up the stairs, on the right." Grabbing my bag I go up the stairs and into the toilet. The light globe is low wattage and there are no windows. In the low light it is still possible to see dark smears on the wash basin and toilet cistern. I take the one tissue I have in my bag and wipe the toilet seat and quickly pee. I then use the tissue to turn on the tap, rinse my hands and turn off the tap. I flush with the tissue covering my finger and throw it into the bowl. I flip the door handle down with my elbow and say under my breath, "It's Ok. It's Ok. I've got no cuts on my arse or hands. The blood can't get in. It can't get in."

I have never publicly shared the story of the toilet-cum-shooting gallery. Publications from the study use survey and interview data only. This is indicative of my publications in the area of injecting drug use which rely on interview and survey data, not field notes. Recently, I have come to look back over field notes from a number of different studies to consider why there are some tales I chose to publicly tell and others which are set aside. I have found that the stories that most trouble me (and which are invariably relegated to silence) are those where my encounter with the Other elicited a tangle of embodied, emotional, ethical and epistemological concerns that sedimented within me over a long period. These have become the "stuck places" of ethnographic research because they refuse easy interpretation and representation and have incomplete ethical answers (Lather, 2001, p. 477).

What troubles me about the Sanday account is not so much the element of risk from either Robbo crashing through the door brandishing a big lump of wood or the remote possibility of becoming infected with Hepatitis C. Unlike other ethnographers who have wisely chosen to share stories of dangers in the field (see Nordstrom & Robben, 1995), I relate this story for a different reason. Its absence is conspicuous in my work on injecting cultures. Technically, the account is a piece of good evidence illustrating the very real risks for the environmental transmission of Hepatitis C. When injecting occurs in places frequented by people who inject drugs (PWID), there is a possibility that someone else's infected blood may be inadvertently (sometimes invisibly) left at the place and transferred to a piece of injecting equipment or injecting site on the body, therefore increasing risk of infection (Alter, 2007; Southgate, et al., 2003). This risk has lead to the development of health promotion messages about blood awareness (Davis & Rhodes, 2004). There is ongoing interest in shooting galleries, usually public or semi-public places where PWID go to have "shot," and environmental factors for the transmission of blood borne viruses (Page et al., 2006; Parkin & Coomber, 2010; Rhodes et al., 2006). Despite this, I have predominantly written about the issue using interview and survey data and rarely referred to field notes (Hopwood et al., 2003; Southgate, et al., 2003, 2005; Southgate & Hopwood, 2001).

My refusal to use field notes is an instructive complication that reveals something about the dynamics of research with marginalised people and their advocates and the production of evidence. A key part of such research involves emotional border crossing (Blackman, 2007; Southgate, 2011). In this case, the emotional border crossing consisted of a commitment to health workers and their client group not to publish accounts that are considered voyeuristic or unsavoury. Part of the compact between researchers and the health workers who facilitate the research is to protect marginalised groups from further demonisation by producing graphic accounts. Even when PWID provide graphic accounts of risk behaviour in interviews, there is always an ethical question about whether to use these accounts, even if they are carefully re-presented in "scientific" explanation, as the effect may be that prejudices are confirmed and the Other further pathologised.

My refusal to use the Sanday account encapsulates the tension between the worth of it as useful evidence and the consequences of betraying the social bond between myself and health workers who helped me do research. While there is a growing literature on emotions and knowledge production in research (Gould, 2010; Lee-Treweek, 2000; Watts, 2008), there are relatively few explicit accounts of emotional decision-making and the production of knowledge in research, particularly the use of field notes (Punch, in press). Jackson (2010) challenges us to:

> examine "what happens" (and/or does not) when the ethnographic sparks start flying in the intersubjective collision that is anthropological research, an anthropology that defines its difference from most other versions of the social scientific enterprise (at least in their overly scientistic iterations) as marking time between a version of ethnography seen as a transparent window into a discrete, passive, and objective social world and a contrasting variety considered something more like a black box of feedback loops, inter/subjective contaminations, and almost unteachable artistry. (p. 281)

This challenge to talk about intersubjective collisions should also include the intersubjective collusions that occur in the production of "useful" evidence (Bourdieu, 1999; Pitt & Britzman, 2003). Discussing the consequences of such collusions serves to open up the "black box," exposing the types of seduction that can characterise ethnographic fieldwork (Robben, 1995, 1996). Of course, these collusions are not necessarily bad in terms of their epistemic and ethical consequences. To paraphrase Foucault (1983), not everything is bad, but everything is dangerous, which is not the same as bad. My refusal to publish the Sanday account before now, some decade after its occurrence and since my departure from the area of public health is a instructive complication that speaks to the situated character of ethical decision making in research and its consequences in the production of knowledge.

This instructive complication is more than a pragmatics of doing research with hidden, marginalised populations: It is easier to do research with PWIDs if you have health workers on-side, although not impossible. It is instructive because it demonstrates an emergent ethics (Parker, 2007) or the creative, intersubjective negotiation of method, interpretation and representation by researcher, the researched, and importantly in this case, by those who facilitate the process. Such a practice is different from adherence to the universal principles of ethical conduct espoused by institutional ethics committees, although it certainly draws on them (Meskell & Pels, 2005). It is an ethics that emerges from practice, that in this case includes a dynamics of collusion and refusal. Acknowledging the epistemic consequences of emergent ethics is important because it represents a refusal of rationalist ideas about the production of evidence, so powerful in areas such as public health.

FIELD 2: MOONLIGHTS BROTHEL

Field diary extract – November 2003

> I can't believe I saw Denise. I was sure it was her. We arrived at Moonlights, a discrete brothel housed in a boring red-brick brick building on a main road. From the front, nothing really says *brothel*. There are no large street numbers on the front of the building. There is no neon 'open' sign or a double entrance, where the clients can slip behind a screen to ring the buzzer for entry. It's the second time I've been here, and my first visit with Sabelle (who only started the peer education job two weeks ago). There is a long hallway running up the centre, with bedrooms off it and a ladies lounge and kitchen where the workers hang out while they wait for clients. It's late afternoon and we are shown to the lounge by the receptionist. There are four women there and they seem pleased that we've arrived. I sit on the lounge and begin to take out information on the sexual health topic of the week when another sex worker enters the lounge and sits down opposite me. I was sure it was Denise as soon as I saw her. Our eyes met for a fraction of a second, she reddened, and lifted the collar of her fluffy pink dressing gown to cover her face. She shifted herself so that her back is to me. The doorbell signals that a client has arrived. He is put in a bedroom. The receptionist calls the name of each woman who then goes to meet the client. When she calls out "Sasha," Denise gets up, shrugs off the dressing gown and strides into the bedroom. I

> decide to pretend that I don't know her. I don't want to make her feel embarrassed. I'm a bit red too. When she returns she picks up the gown, turns and says, "Erica, is that you? Oh my god you look different." I get up and hug her saying, "Don't you say that I've got fat." We laugh, and then she breaks away and she goes to the kitchen which is out of sight. She doesn't return for the education session. I don't see her again. We leave.

This account exemplifies the process of "getting lost" in research (Lather, 2007). It is not a story about the pathway to the transcendental evidence of truth and reality (Koro-Ljungberg, Yendol-Hoppey, Smith, & Haues, 2009). Denise was the ex-partner of a friend of mine. I had not seen her for a couple of years before this encounter and I did not know she was a sex worker. While the encounter was an example of the emotionally awkward things that can happen when we do research (Southgate, 2011), it is also emblematic of "difficult knowledge" where there is a breakdown in interpreting and representing experience in research (Britzman, 1998; Lather, 2007).

Research, to use a phrase from Sedgwick (2008), is peppered with the "bathos of unknowing" (p. 77). It is just that researchers rarely talk about it. The act of collecting data, interpreting it and re-presenting it, can veer between moments of extreme clarity and occasions when we admit that we just don't know what exactly happened, what it meant and whether or not it is useful. There is professional embarrassment in coming across something that defies our understanding. Our impulse is to forget about it, discard it from the useful data pile, or quietly return to it, over and over again, in order to try to understand, usually without success. To admit to not-knowing carries a "negative moral load" (Dilley, 2010, p. 179).

My encounter with Denise has produced a vacillation between knowing a bit and not knowing at all (Lather, 2003). I know how I felt at the time – embarrassed because my presence may have made Denise embarrassed. However, it would be dishonest to suggest that I actually understood how Denise felt and a conceit to produce an interpretive account that linked these feelings to her actions and intent. I could read the reddening of Denise's face or the pulling up of her collar as signs of shock or shame as a result of the stigma of being found out as a sex worker, a spoiled identity in Goffman's (1968) parlance. But this would be an example of heavy-handed, intrusive analysis resulting from an expectation that researchers need to produce some meaning from their data (Wolcott, 2009, p. 33).

Even within constructivist and critical paradigms where positivist notions of validity, reliability and objectivity have been superseded by ideas of credibility, dependability and transferability in data analysis (Denzin & Lincoln, 2005), there are some research occasions when it is impossible to understand what is happening and what it means. Despite the inclinations of institutional ethics committees and others within the social sciences to insist on practices such as member checking (Flick, 2007; Hammersley & Atkinson, 1995; Lincoln & Guba, 1985), it is not always possible to go back to participants to have them validate your interpretation. Sometimes it may not be ethical either. After my encounter with Denise I made a decision, based on a gut feeling, not to go back to Moonlights. I thought through a number of worst case scenarios: If I returned she might become distressed, feel the need to hide when I visited or even feel as though she should work elsewhere for fear of seeing me again. I didn't know if this would be the case but it was part of an emergent ethical decision-making process and this precluded finding out the "truth" of the encounter from Denise's perspective.

I am reminded of another occasion during the same research project when a very angry massage parlour manager ordered the health worker and me off the premises while we were conducting an education session with the workers. The manager began ranting at us to get out but I could not take in what exactly she was saying. The health worker, maintaining her dignity responded by saying, "I will not be spoken to like that. We are leaving." In the car afterwards, we were both at a loss to explain what occurred, and neither of us knew what had provoked the incident. The only thing we did know was that we had tried to leave with some dignity intact.

Last (1981), in his classic article, "The Importance of Knowing about Not-Knowing," observes that the "reluctance in ethnography to record what people do not know is understandable: it is hard enough to record what they do know" (p. 387). I would argue that this reluctance also applies to researchers. Occasions of not knowing, moments when something happened that defied easy interpretation then and afterwards, are instructive because they make us give up our attachment to the "lovely knowledge" (Pitt & Britzman, 2003, p. 766) of research or the "right" interpretation, its core truth and implications. Not knowing is different from not-caring-to-know (Last, 1981) or an indifference to data they may not fit the theories or interpretative frameworks we favour, or is annoying in its contradictions (for a fascinating account of unruly data see Power, 2004). Not-knowing is also different from knowing-but-not-telling, the dynamic evident in the Sanday account. Every empirical research project is littered with the detritus of occasions of not-knowing, not-caring-to-know, and knowing-but-not-telling. Spotlighting these moments is as instructive as producing lean, mean sanitised research findings (Huisman, 2008).

FIELD 3: POSTERMERE PRIMARY SCHOOL

Research diary extract – March 2011

> I just met with Alison, the principal of Postermere Primary School which is in a suburb characterised by social housing and high unemployment. Alison had been told about a research project on engaging community in low socioeconomic school communities. It's about 8am and she takes me for a wander through the office block to meet one of the mums who does the breakfast club. She's making toast. A couple of older kids are helping out. We stroll through the playground and Alison points out some play equipment that has recently been installed courtesy of some new funding. She explains that the playground was pretty bare before and that a lack of equipment sometimes caused fights between students. Alison wants to engage parents

> more. I ask her what she means by engagement. She thinks for a few seconds: "You know, to get them in here, to feel comfortable with the school and to understand the right way to behave when they come here." I must have looked a bit confused. She goes on to say, "Some parents swear or abuse each other and we can't have that here. We want you to help us change the launch pad." "Launch pad?" I ask. Alison responds, "You know, where the kids come from. We want you to change the launch pad so the kids don't come here hungry, tired and dirty. So we don't have to spend the morning dealing with all the other issues instead of learning. You know, the launch pad that the kids come from." I tell her that it might be beyond my capabilities to address all the issues of the launch pad but we can think about some ways to engage the community in student learning. I walked away feeling pretty scared about Alison's expectations.

I have just started research into the area of engaging community in low SES school communities. The research is driven by national and state education policy agendas that view community (usually parental/carer) engagement as key to fostering the educational and career aspirations of this group. The area is mired in assumptions – that community partnerships are key to lifting student achievement; that these parents and carers are less engaged or able to engage than their higher SES peers; that teachers in these schools should and can do better at forging community connections; and that students from low SES backgrounds should aspire to tertiary, and particularly University, education as a norm. While there are various evidence bases for these assumptions (some of which are more convincing than others), the main problem with the idea of engaging community is not so much the concept of reaching out (for this would be hard to argue against), but rather the idea that policy and its attached funding opportunities are somehow reflections of an innocent phenomenon that exists "out there." As Bacchi (2000) suggests:

> (I)t is inappropriate to see governments as responding to 'problems' that exist 'out there' in the community. Rather 'problems' are 'created' or 'given shape' in the very policy proposals that are offered as 'responses.' (p. 48)

The area of mass education has long been flavoured by a redemptive promise. The redemptive promise of education is its potential to bring economic development, social efficiency and justice and increased social mobility by severing the link between social origin and destination (Dale, 2007). This redemptive promise is reflected in educational research, which is often premised on delivering social transformation (Lee, 2010; McWilliam & Lee, 2006). McWilliam and Lee (2006) argue that the redemptive hope is a "fantasy," not because we should dismiss the desire for a new and better social order achieved through education, but because of the "seductiveness and elusiveness of that hope" (p. 46).

The difficulty is in living with the seduction and elusiveness of the hope that educational research can fix social problems, particularly when those problems are preordained in public policy which charges researchers and practitioners with fixing them. Alison's desire to "change the launch pad" or the social milieu in which her students live, demonstrates both an accurate reading and a misreading about what educational research, and the evidence which stems from it, should and can do. It is an accurate reading because, from a practitioner and policy perspective, the true goal of such educational research is to enact the redemptive promise. As McWilliam and Lee (2006) point out, it is probably not wise or possible to dismiss the collective or individual hope for a new and better social order through education. At the same time, they also suggest that it is an "extravagant promise." The issue with extravagant promises is that its advocates, practitioners and researchers (but arguably not policymakers), are held responsible when they fail to deliver (Connell, 2009). Increasing social equity and inclusion through education is a "wicked problem" (Rittel & Webber, 1973), one which can be variously defined by any number of stakeholders and is resistant to easy, generalisable solutions.

Even after years of researching wicked social problems, I find them anxiety-provoking, not just because they are hard to grapple with, resistant to research methodologies, and materially and ethically complex. My main anxiety stems from how to have conversations about what evidence is, how it is produced and the limits of its extravagant promise to bringing about change. These conversations are somewhat easier between researchers than they are with policymakers or practitioners. The question for me is how to have an honest conversation, particularly with practitioners like Alison, which simultaneously buys into the redemptive project and refuses (some of) it. It is a conversation that must include a dialogue on the epistemic binary logic: The truth is out there/There is no "truth" (Fusco, 2008). Occasions like my meeting with Alison reminded me that my happy positivism is a sunny epistemic disposition that precludes such a conversation. This is a dynamics of not-speaking-of-knowing. How do I tell Alison that I will try, like the good researcher I am, to make up some useful evidence, but that I can't promise to be able to produce something that will transform something as big and as complicated as the "launch pad?"

THE BENEFITS OF TAKING A MOMENT

Koro-Ljungberg, et al.(2009) call for greater epistemological awareness in qualitative research. They suggest that this involves exploring the conditions under which knowledge and "truths" are produced, justifications for these, and instantiation. In this chapter I have set about describing three research occasions in which ethical and intersubjective dynamics produce a set of conditions which had epistemic consequences. Specifically, I describe some of the conditions that produce knowing-but-not-telling, not-knowing, and not-speaking-of-knowing. These types of knowing are linked by an emergent ethics, played out in what Fine (1998) describes as the intersubjective hyphen between researcher-other:

> I mean to invite researchers to see how these 'relations between' get us 'better' data, limit what we feel free to say, expand our minds and constrict

> our mouths, engage us in intimacy and seduce us into complicity, make us quick to interpret and hesitant to write. (p. 135)

In this chapter I have concentrated on the fairly rare occasions over a ten year period when I experienced a heightened epistemological awareness. The focus on occasions fits a version of poststructural feminism that is concerned with the empirical, the material and the intersubjective (Hojgaard & Sondergaard, 2011). In line with this, I would contend that there is a need to spotlight what Orner, Ellsworth and Miller (1996) describe as "excessive moments," or those occasions during research that cannot be contained and tamed by accepted research protocols, easy categories of interpretation, realist tales, or simple, predictable ethical arrangements.

This spotlighting strategy is both consistent with but different from the approach of poststructural feminists like Lather (2007) who desire a perpetual double movement where feminist researchers must continually work within and against the foundational categories of their science. I must admit that I find the idea and practice of a perpetual double movement, neither appealing nor practical. It is not the idea of critique that I dislike but the perpetual motion. I object to being in a constant state of ontological and epistemological double movement. It would feel a bit like using one of those air rider walking machines where you get exercise benefits from striding back and forth (in a double movement) on the same spot but don't really get to see anything interesting. Instead, I prefer to wander happily around undertaking empirical research, and (usually) writing up realist tales of what I have found and what it might mean for practice and policy. Perhaps, this approach to research is an example of Lather's (2007, p. 164) musing on what will come after postpositivism, when she cites the possibilities "neopositivist" or "neopragmatist" approaches to theory and research.

To be a happy positivist, in the Foucauldian sense, is not to unquestioningly accept that everything is knowable and that the truths are objective or neutral. Rather it is to take a "stone-cutting" (Foucault, 1977, p. 140) approach to what we do as researchers by paying careful attention to the details of the material and empirical, the exact moments when we cannot produce anything like correct knowledge or secure understanding. It is in the casting of light on exact occasions, points in time and (intersubjective) space, that the instructive complications of knowledge production can be seriously examined. There are great epistemological benefits in occasionally taking a moment.

NOTE

[i] In order to ensure anonymity, pseudonyms have been used for the names of people, places and organisations.

REFERENCES

Alter, M.J. (2007). Epidemiology of hepatitis C virus infection. *World Journal of Gastroenterology, 13*(17), 2436-2441.

Bacchi, C. (2000). Policy as discourse: What does it mean? Where does it get us? *Discourse: Studies in the Cultural Politics of Education, 21*(1), 45-57.

Bacchi, C. (2010). *Foucault, policy and rule: Challenging the problem-solving paradigm*. Aalborg: FREIA – Feminist Research Center in Aalborg.

Blackman, S.J. (2007). 'Hidden ethnography': Crossing emotional borders in qualitative accounts of young people's lives. *Sociology, 41*(4), 699-716.

Bourdieu, P. (1999). Understanding. In P. Bourdieu (Ed.), *The weight of the world: Social suffering in contemporary society*. (pp. 607-626). Cambridge: Polity Press.

Britzman, D. (1998). *Lost subjects, contested objects: Towards a psychoanaltic inquiry of learning*. New York: SUNY.

Clegg, S. (2010). The possibilities of sustaining critical intellectual work under regimes of evidence, audit, and Ethical governance. *Journal of Curriculum Theorizing, 26*(3), 21-35.

Connell, R. (2009). Good teachers on dangerous ground: Towards a new view of teacher quality and professionalism. *Critical Studies in Education, 50*(3), 213-229.

Dale, R. (2007). Educating. In S. Mathewmann, C.L. West-Newman, & B. Curtis (Eds.), *Being sociological* (pp. 295-315). Houndmills, Basingstoke: Palgrave MacMillan.

Davis, M., & Rhodes, T. (2004). Managing seen and unseen blood associated with drug injecting: Implications for theorising harm reduction for viral risk *International Journal of Drug Policy, 15*(5-6), 377-384.

Denzin, N.K., & Lincoln, Y.S. (2005). Introduction: The discipline and practice of qualitative research. In N.K. Denzin & Y.S. Lincoln (Eds.), *The Sage handbook of qualitative research* (Third ed.). Thousand Oaks, CA: Sage.

Dilley, R. (2010). Reflections on knowledge practices and the problem of ignorance. *Journal of the Royal Anthropological Institute, 16*, 176-192.

Eisenhart, M., & Towne, L. (2003). Contestation and change in national policy on "scientifically based" education research. *American Educational Research Association, 32*(7), 31-38.

Fine, M. (1998). Working the hyphens: Reinventing self and other in qualitative research. In N. Denzin & Y.S. Lincoln (Eds.), *The landscape of qualitative research: Theories and issues* (pp. 130-155). Thousand Oaks, CA: Sage.

Flick, U. (2007). *An introduction to qualitative research* (4th ed.). London: Sage.

Foucault, M. (1972). *The archaeology of knowledge and the discourse on language*. New York: Pantheon Books.

Foucault, M. (1977). *Discipline and punish: The birth of the prison*. London: Penguin Books.

Foucault, M. (1983). On the genealogy of ethics. In H.L. Dreyfus & P. Rabinow (Eds.), *Michel Foucault: Beyond structuralism and hermeneutics* (2nd ed.). New York: The New Press.

Foucault, M. (2003). So is it important to think? In P. Robinow & N. Rose (Eds.), *The essential Foucault*. New York: The New Press.

Fusco, C. (2008). 'Naked truths': Ethnographic dilemmas of doing research on the body in social spaces In K. Gallagher (Ed.), *The methodological dilemma: Critical, creative, and post-positivist approaches to qualitative research* (pp. 159-184). London/ New York: Routledge.

Goffman, I. (1968). *Stigma: Notes on the management of spoiled identity*. Harmondsworth: Penguin.

Gould, K.A. (2010). Anxiety, epistemology, and policy research "behind enemy lines." *Geoforum, 41*(1), 15-18.

Hammersley, M. (2007). The issue of quality in qualitative research. *International Journal of Research and Method in Education, 30*(3), 287-305.

Hammersley, M. (2008). Troubling criteria: A critical commentary on Furlong and Oancea's framework for assessing educational research. *British Educational Research Journal, 34*(6), 747-762.

Hammersley, M., & Atkinson, P. (1995). *Ethnography: Principles in practice* (2nd ed.). London: Routledge.

Hojgaard, L., & Sondergaard, D.M. (2011). Theorizing the complexities of discursive and material subjectivity: Agential realism and poststructural analysis. *Theory and Psychology, 21*(3), 338-354.

Hopwood, M., Southgate, E., Kippax, S., Bammer, G., Isaac-Toua, G., & MacDonald, M. (2003). The injection of methadone syrup in New South Wales: Patterns of use and increased harm after partial banning of injecting equipment. *Australian and New Zealand Journal of Public Health, 27*(5), 551-555.

Huisman, K. (2008). "Does this mean you're not going to come visit me anymore?": An inquiry into an ethics of reciprocity and positionality in feminist ethnographic research. *Sociological Enquiry, 78*(3), 372-396.

Israel, B.A., Schulz, A.J., Parker, E.A., & Becker, A.B. (1998). Review of community-based research: Assessing partnership approaches to improve public health. *Annual Review of Public Health, 19*, 173-202.

Jackson, J. L. (2010). On ethnographic sincerity. *Current Anthropology, 51*(Supplement 2), S279-S287.

Koro-Ljungberg, M. (2010). Validity, responsibility, and aporia. *Qualitative Inquiry, 16*(8), 603-610.

Koro-Ljungberg, M., Yendol-Hoppey, D., Smith, J.J., & Haues, S.B. (2009). (E)pistemological awareness, instantiation of methods, and uninformed methodological ambiguity in qualitative research projects. *Educational Researcher, 38*(9), 697-699.

Last, M. (1981). The importance of knowing about not-knowing. *Social Science and Medicine, 15B*(3), 387-392.

Lather, P. (2001). Postmodernism, post-structuralism and post(critical) ethnography: Of ruins, aporias and angels. In P. Atkinson, A. Coffey, S. Delamont, J. Lofland, & L. Lofland (Eds.), *Handbook of ethnography* (pp. 477-487). London: Sage.

Lather, P. (2003). Applied Derrida: (Mis)Reading the work of mourning in educational research. *Educational Philosophy and Theory, 35*(3), 257-270.

Lather, P. (2006). Paradigm proliferation as a good thing to think with: Teaching research in education as a wild profusion. *International Journal of Qualitative Studies in Education, 19*(1), 37-57.

Lather, P. (2007). *Getting lost: Feminist efforts towards a double(d) science.* New York: SUNY.

Lather, P. (2009). Against empathy, voice and authenticity. In A.Y. Jackson & L.A. Mazzei (Eds.), *Voice in qualitative inquiry: Challenging conventional intepretive and critical conceptions in qualitative research.* London/ New York: Routledge.

Lee-Treweek, G. (2000). The insight of emotional danger: Research experience in a home for older people. In G. Lee-Treweek, & S. Linkogle (Eds.), *Danger in the field: Risk and ethics in social research.* London: Routlege.

Lee, A. (2010). What counts as educational research? Spaces, boundaries and alliances. *The Australian Educational Researcher, 37*(4), 63-78.

Lincoln, Y.S., & Guba, E.G. (1985). *Naturalistic inquiry.* Beverly Hill, CA: Sage.

Mazzei, L.A. (2007). *Inhabited silence in qualitative research: Putting poststructural theory to work.* New York: Peter Lang.

McWilliam, E., & Lee, A. (2006). The problem with 'the problem with educational research.' *The Australian Educational Researcher, 33*(2), 43-60.

Meskell, M., & Pels, P. (2005). Introduction: Embedding ethics. In M. Meskell, & P. Pels (Eds.), *Embedding ethics: Shifting boundaries of the anthropological profession* (pp. 29-46). Oxford: Berg Publishers.

Nagy Hesse-Biber, S. (Ed.). (2011). *Handbook of feminist research* (2nd ed.). Thousand Oaks, CA: Sage.

Nordstrom, C., & Robben, A. (Eds.). (1995). *Fieldwork under fire: Contemporary studies of violence and survival.* Berkeley: University of California Press.

Orner, M., Miller, J.L., & Ellsworth, E. (1996). Excessive moments and educational discourses that try to contain them. *Educational Theory, 46*(1), 71-91.

Page, J.B., Shapshak, P., Duran, E.M., Even, G., Moleon-Borodowski, I., & Llanusa-Cestero, R. (2006). Detection of HIV-1 in injection paraphernalia: Risk in an era of heightened awareness. *AIDS Patient Care and STDs, 20*(8), 576-585.

Parker, M. (2007). Ethnography/ethics. *Social Science and Medicine, 65*(11), 2248-2259.

Parkin, S., & Coomber, R. (2010). Fluorescent blue lights, injecting drug use and related health risk in public conveniences: Findings from a qualitative study of micro-injecting environments. *Health and Place, 16*(4), 629-637.

Pitt, A., & Britzman, D. (2003). Speculations on qualities of difficult knowledge in teaching and learning: An experiment in psychoanalytic research *International Journal of Qualitative Studies in Education, 16*(6), 755-776.

Power, E.M. (2004). Toward understanding in postmodern interview analysis: Interpreting the contradictory remarks of a research participant. *Qualtative Health Research, 14*(6), 858-865.

Punch, S. (in press). Hidden struggles of fieldwork: Exploring the role and use of field diaries. *Emotion, Space and Society.*

Reed-Danahay, D. (1997). *Auto/ethnography.* New York: Berg.

Rhodes, T., Kimber, J., Small, W., Fitzgerald, J., Kerr, T., Hickman, M., et al. (2006). Public injecting and the need for 'safer environment interventions' in the reduction of drug-related harm. *Addiction, 101*(10), 1384-1393.

Rittel, H., & Webber, M. (1973). Dilemmas in a general theory of planning. *Policy Sciences, 4*(2), 155-169.

Robben, A. (1995). The politics of truth and emotion among victims and perpetrators of violence. In C. Nordstrom & A. Robben (Eds.), *Fieldwork under fire: Comtemporay studies of violence and survival.* Berkely: University of California Press.

Robben, A. (1996). Ethnographic seduction, transference, and resistance in dialogues about terror and violence in Argentina. *Ethos, 24*(1), 71-106.

Sedgwick, E.K. (2008). *Epistemology of the closet.* Berkeley: University of California Press.

Southgate, E. (2003a). Liquid handcuffs: A tale of power, subjectivity and risk and the drug treatment clinic. In M. Tamboukou, & S. Ball (Eds.), *Genealogy and ethnography: Fruitful encounters and dangerous liaisons.* New York and London: Peter Lang.

Southgate, E. (2003b). *Remembering school: Power, subjectivity and emotion in stories of school life.* New York/London: Peter Lang.

Southgate, E. (2011). Decidedly visceral moments: Emotion, embodiment and the social bond in ethnographic fieldwork. *International Journal of Work Organisation and Emotion, 4*(3/4), 236-250.

Southgate, E., Day, C., Kimber, J., Weatherall, A.M., MacDonald, M., Woolcock, G., et al. (2003). *Dealing with risk: A multidisciplinary study of injecting drug use, Hepatitis C and other blood-borne viruses in Australia.* Canberra: Australian National Council on Drugs.

Southgate, E., & Hopwood, M. (2001). The role of folk pharmacology and lay experts in harm reduction: Sydney gay drug using networks *Australian and New Zealand Journal of Public Health, 27*(5), 551-555.

Southgate, E., Weatherall, M., Day, C., & Dolan, K. (2005). What's in a virus? Folk understandings of Hepatitis C infection and infectiousness among injecting drug users in Kings Cross, Sydney. *International Journal for Equity in Health, 4*(5), 177-189.

Suri, H. (2008). Ethical considerations in synthesising research: Whose representations? *Qualitative Research Journal, 8*(1), 62-73.

Trotter, G. (2003). Buffalo eyes: A take on the global HIV epidemic. *Cambridge Quarterly of Heathcare Ethics, 12*(4), 434-443.

Van Maanen, J. (1988). *Tales of the field: On writing ethnography.* Chicago: University of Chicago Press.

Watts, J.H. (2008). Emotion, empathy and exit: Reflections on doing ethnographic qualitative research on sensitive topics. *Medical Sociology Online, 3*(2), 3-14.

White, J., & Drew, S. (2011). Colecting data or creating meaning? Troubling athenticity in ethnographic research. *Qualitative Research Journal, 11*(1), 3-12.

Wolcott, H.F. (2009). *Writing up qualitative research.* Los Angeles: Sage.

Woodthorpe, K. (2009). Reflecting on death: The emotionality of the research encounter. *Mortality, 14*(1), 70-86.

Erica Southgate
School of Education
University of Newcastle, Australia

THI TUYET (JUNE) TRAN AND JULIE WHITE

MANAGED UNIVERSITIES

Vietnam and the West

INTRODUCTION

In this essay significant and recent changes in Western higher education provide a useful backdrop for this examination of higher education in Vietnam, a developing, challenged and complex nation. The main argument forwarded here is that Vietnamese higher education is externally managed and that this equates to hegemony. By bringing together different literatures the extent of how academics and the curriculum are 'managed' in both contexts is revealed.

Although Vietnam is an independent country, the policies and influence driving the system come from outside, mostly from Western developed countries. Globalization and internationalization are often used to explain and justify the tendency of the Vietnamese policy-makers who repeatedly look to the West (Phuong-Mai Nguyen, Elliott, Terlouw, & Pilot, 2009). Implicit in this practice is a naïve belief that by adopting the successful policies of the developed world, higher education in Vietnam will develop, firstly, more quickly; secondly, it will readily integrate into the international academic world; and thirdly, that Vietnamese higher education will structure its programs in order to attain international standards. In this sense, globalization offers the mechanism and justification for the diffusion of Western ideas to the point that these have become universal and dominant in the academic world (Rivzi, 2007). By adopting and adapting to the Western policies, the Vietnamese Higher Education System (HES) depends upon external power and control and has become 'managed.' Postcolonial theory, with its inherent concepts of power, influence, dependency, local identity, difference and resistance provides a useful and productive lens through which to examine Vietnamese higher education. It also serves to raise awareness of implications and pose a warning for Vietnam and other developing countries of the dangers of not only being managed, but also of denying, devaluing and decommissioning local educational histories, cultural values and systems.

THE MANAGED UNIVERSITIES

The contemporary corporatized university is increasingly implementing audit and quality agendas that involve the management of knowledge, curriculum and the work of academics. A significant body of research suggests that universities in many countries are reforming into corporate entities in order to satisfy international policy demands (Johnson, Hirt, & Hoba, 2011) and our focus on Vietnamese

M. Vicars et al. (eds.), Discourse, Power, and Resistance Down Under, 151–160.

higher education adds to this discussion.

Within the Australian context, significant changes have occurred in the recent past, but with the implementation of the Excellence in Research for Australia (ERA) exercise, this change is occurring at an exponential rate. Performativity (Lyotard, 1984) and new managerialism (Deem and Brehony, 2005) are helpful concepts to understand these changes, which have closely followed the lead of England and New Zealand. Almost ten years ago, Blackmore (2003, p. 4) argued:

> Within universities, what counts as knowledge, how knowledge is constituted, and indeed how it is counted, has also been revised by the entrepreneurial or performative state, seeking not to invest in, but rather to accrue wealth from education. The role of universities has shifted more towards managing knowledge production for the state in the national interest rather than the production of disciplinary value for its intrinsic value.

These changes in knowledge production and the sanctioned knowledge taught in universities (White, 2012b), has intensified considerably in Australia as well as in other Western nations. In the Vietnamese context, the extent of the change is even more pronounced. Within its highly centralized higher education system universities do not have much autonomy as the Vietnamese Ministry of Education and Training (MOET) has assumed the power to make most curriculum decisions including what knowledge is of most value, and what the system must teach (Tran Thi Tuyet, 2010a). However, given the desire to increase the quality of the Vietnamese HES to attain international standards (Pham Quang Nghi, 2010), together with the financial support and 'expert' help from developed countries, Western policies and teaching materials have gradually been introduced into the Vietnamese HES. MOET has even funded the direct translation of English language textbooks into Vietnamese.[i] As a result, many textbooks in most disciplines have been translated into Vietnamese without local consideration. Some Vietnamese universities are increasingly adopting the uncritical practice of ordering textbooks from the UK, US or Australia, which are used to determine the curriculum in Vietnam.[ii] Teachers simply learn the knowledge provided in the textbook and 'transfer' it to their students (Hoang Tuy, 2009). Vietnamese academics do not contribute to the production of knowledge, and have been positioned into more technical roles, with the Vietnamese universities increasingly only transferring knowledge, in many cases, from the West that does not necessarily meet national interests.

In Australia and other Western nations, curriculum in higher education is increasingly controlled externally through accreditation and standards regimes (White, 2012b) and demands for changes to pedagogy in higher education have been well documented (Lee, 2005; Strathern, 1997; Davies et al., 2005; McWilliam, 2004). Processes of audit, managerialism and performativity have led to less autonomy for individual academics and university departments (Marginson et al., 2009; Rizvi & Lingard, 2010). Within the Vietnamese context, university curriculum and pedagogy are also managed, but the management process is more extensive than seen in the West. University lecturers are obliged to follow the curriculum frameworks established by MOET, where external standards and accreditation processes have a powerful influence. As well as managing university curriculum, university teacher pedagogy is also influenced and shaped by external requirements. Some of the 'advanced' and 'modern,' and therefore superior, Western approaches to learning now required in Vietnamese higher education are student-centred learning, credit-based learning, cooperative learning and autonomous learning (Phuong-Mai Nguyen, Elliott, Terlouw, & Pilot, 2009; Tran Thi Tuyet, 2010b). Demands for change in pedagogy from a highly teacher-centred teaching to student-centred learning in higher education in Vietnam are also well documented and explicitly announced to the entire HES (Pham Thanh Nghi, 2010). Teachers do not appear to have any autonomy over what to teach and how to teach their students.[iii]

Increasingly academics in the West have become 'managed professionals' (Codd, 2005) and few can escape this management of their work (Blackmore, 2003). The ways in which professional identity is performed in Western nations in the traditional areas of research, teaching and service have been addressed elsewhere (White, 2012a) and who can forget the memorable 'Weasel' character in Sparkes' (2007) exploration of managed academic work? Being a managed academic in the West does not appear to produce many positive outcomes, but in the Vietnamese higher education context, the situation is more extreme. University teachers are underpaid and they overwork. Their main focus is teaching, rather than undertaking disciplinary service or research (Hayden & Lam Quang Thiep, 2010). The prevailing image of the Vietnamese academic (in Vietnam) is the 'teaching only' employee who has one leg longer than the other – with the outside leg longer than the inside one, to enable them to moonlight, by teaching long hours in order to earn an adequate income (Harman & Nguyen Thi Bich Ngoc, 2010). There is limited funding for research in Vietnam and there has been a traditional focus on teaching rather than research. Coupled with the high workloads and low payment of university lecturers, this indicates a very dark picture of research potential in Vietnamese higher education. The 'Western privilege' (Phuong-Mai Nguyen, Terlouw, & Pilot, 2005) over contemporary Vietnamese policy and advanced and 'better' teaching pedagogies, ties Vietnamese higher education to the values of Western nations.

GLOBALIZATION

Globalization and internationalization are often referred to as justification for the tendency of Vietnam to continually looking westward for direction in policy and practice. This appears to be occurring in many developing countries, so Vietnam is not an exception. It has also been suggested that current policies and practices in higher education in Vietnam (and elsewhere) cannot be fully understood without placing it in its national and global context (Apple, 2005). Massification of higher education participation has succeeded in Vietnam with the number of students enrolled in the system increasing 13 times in 2009 compared to the number enrolled in 1987 (The Vietnamese General Statistics Office, 2009). Vietnam is under pressure to reform and modernize its educational system in order to attain international standards set by Western developed nations (Pham Lan Huong & Fry, 2004). The wholesale and uncritical adoption of Western policies and practices, regarded as more advanced and effective (Deem, Mok, & Lucas, 2008, p. 21), are systematically replacing Vietnamese higher education policies and practices.

The adoption of Western policy into Vietnamese higher education is also part of the compliance requirements involved in the acceptance of capitalization assistance from the developed world to the developing world (Altbach, 2004). This ensures that Western policy and practices are adopted as Vietnamese authorities lack research capability to develop their own policies and practices (Harman & Nguyen Thi Bich Ngoc, 2010).

Globalization has become a powerful force in driving higher education in Vietnam and related phenomenon such as the development of information technology and the imbalance of power and influence in the world, where the Majority World – the less developed world – has increasingly depended on the Minority World – the developed world (Singh, Kenway, & Apple, 2005). When the Minority world spends a significant amount of money on scholarship and training programs, textbook translations, student exchanges and institution building, this works to further impact on the international academic community (Altbach, 2004), by increasing the dependence of the ones who receive offers (the Majority world) on the ones who supply the offers (the Minority world). Academics are therefore further positioned and managed. Current debates about the way that society, under the globalization process, is organized around flows, such as 'flows of capital, flows of information, flows of technology, flows of organizational interaction and flows of images, sounds and symbols' (Castells, 2000, p. 442). As Rizvi observes (2004, p. 159), these flows are largely uni-directional 'from "the West" to "the rest."' Rizvi has been critical about the relationship between globalization and education and has challenged the encompassing nature of globalization and its explanatory power, particularly the tendency to use globalization to describe any and every aspect of contemporary life by arguing that:

> much of the recent theorization of globalization assumes it (globalization) to be an objective self-evident entity, and does not attend sufficiently to the task of historicizing it, pointing to the hegemonic role it plays in organizing a particular way of interpreting the world. (2007, p. 256)

Rizvi questions the nature of the relationship between the global context and educational change, where he points to the need of historicizing globalization and also to the hegemonic role it has played in the global context in general. Vietnamese higher education seems to have fallen into the trap of globalization rhetoric and hegemony. Rizvi suggests that there is a need to 'understand contemporary ideological constructions of globalization historically' (p. 529). Postcolonial theory, with its animated concepts of place, identity, modes of resistance and difference, is useful in considerations of globalization and the relationship between globalization and education in contemporary world order.

POSTCOLONIAL THEORY AND NEOCOLONIALISM

Postcolonial theory, with its insistence on the cultural dimensions of imperialism and colonialism, helps to explain the acceptance and adoption of neoliberal ideas, which appear as 'natural and inevitable response to the steering logic of economic globalization' (Rizvi, 2007, p. 259). Postcolonial theory also brings globalization back to its imperialist origins. It explains power imbalances and suggests an indirect form of control in the world through economic and cultural dependence. Neocolonialism, a term used within postcolonial theory, is used to point to this imperialism (Ashcroft, Griffiths, & Tiffin, 1998).

Neocolonialism typically refers to the 'control of states by external powers despite the formal appearance of constitutional independence' (Bray, 1993, p. 334). Altback and Kelly (1978, p. 30) suggests that neo-colonialism 'constitutes the deliberate policies of the industrialized nations to maintain their domination.' It may function through foreign-aid programs, technical advisers, publishing firms and via such international agencies as the World Bank (WB), United Nations Educational, Scientific and Cultural Organization (UNESCO), Asian Development Bank (ADP) or the Organization for Economic Cooperation and Development (OECD). In order to qualify for loans or other forms of educational or economic aid, many developing countries struggle to satisfy the loan conditions of the Western developed countries. Indeed, colonialism 'sought to bring communities across vast distances into a single political space, controlled and coordinated from a centre' (Rizvi & Lingard, 2010, p. 25). It is partly a planned policy of advanced and developed nations to maintain their influence in developing countries (Albach, 1971). Kwame Nkrumah (1965, p. 16), a post-independence Ghanaian leader, suggests that:

> The essence of neo-colonialism is that the State which is subject to it is, in theory, independent and has all the outward trappings in international sovereignty. In reality its economic system and thus political policy is directed from outside.

Education has also become an instrument to 'inscribe the colonizing ideology' (Chilisa, 2005, p. 660) onto the colonized or the developing countries (Albach & Kelly, 1978; Watson, 1994). In other words, the term "neocolonialism," as part of postcolonial theory, describes the continued control of territories by the Western metropolis, without the formal occupation in colonized countries.

However, it is argued that education in colonized settings tends to bring more harm than good for existing local educational systems (Bray, 1993). When the structure of school systems, textbook, curriculum, assessment and pedagogical practices are imported from colonial systems (Watson, 1994), they typically ignore and replace local culture and discourses (Wickens & Sandlin, 2007). Moreover, the education agenda of the colonizers has largely been 'framed, constructed and driven by an ideology aimed at colonizing the mind and alienating the self and creating an individual that did not believe in her/himself' (Chilisa, 2005, p. 660). The managed Vietnamese HES has increasingly become dominated by the hegemonic West, particularly in terms of educational theory, pedagogy and curriculum.

NEOCOLONIALISM AND VIETNAMESE HIGHER EDUCATION

Since the collapse of the old Soviet system, Vietnam has been struggling to find an appropriate way to lead its educational system forward. The superior role of the Western developed countries such as US, UK and Australia, and their success in both economic and educational development has raised Vietnam's desire to learn from their models in order to enhance the sustainable development of both the Vietnamese internal economy and its HES. Since Vietnam applied its open door policy involving the adoption of the market driven economy, the Vietnamese HES has adopted many Western policies such as student-centre learning, autonomous learning, credit based curriculum and teaching.

However, after more than a decade adopting these non-traditional policies and practices, despite much effort on the part of teachers and students in the system and a huge investment of money, positive reports on the application of these outcomes are rare. The adoption and importation of educational policies, theories, pedagogical practices and textbook-based curricula, based on Western values, has been widely criticized in the scholarly literature.

Firstly, this has been illustrated to be inappropriate 'cross-cultural cloning,' since adopting policies across cultures without recognizing the distinctive social and cultural dimensions runs the risk of 'false universalism' (Rose & Mackenzie, 1991). What makes the transfer unsuccessful, it is also argued, relates to the neglect of 'detailed consideration of particular aspects of the culture and heritage of the originating country' (Phuong-Mai Nguyen et al., 2009, p. 109). Some scholars use the culture dimensions developed by Trompenaars and Hampden-Turner (1997) and by Hofstede (1983) to illustrate the inappropriateness of adopting such pedagogies as cooperative learning and student-centred learning into Vietnamese contexts (Phuong-Mai Nguyen et al., 2005; Tran Thi Tuyet, 2010b). In the adopted pedagogies, the teacher is required not to transmit knowledge to students, but instead to act as a facilitator in student learning. Students are expected to be active, to raise their voices and to be critical of the knowledge they receive, despite learning throughout school in more traditional ways. This does not seem to be appropriate in the Vietnamese Confucian-influenced culture where the traditional relationship between teacher and student is still very much alive (Harman & Nguyen Thi Ngoc Bich, 2010), and the face saving culture has remained very strong.

Secondly, the successful application of contemporary Western-inspired policy requires pre-conditions and infrastructure but Vietnam does not seem to have adequately prepared for this. The adoption of a credit-based system based on the flexible study options including student choice of elective subjects and flexible class attendance and delivery mode, provides significant challenges for higher education in Vietnam. Thus, the 'credit learning system has not facilitated new forms of instruction as envisaged' (Harman & Nguyen Thi Ngoc Bich, 2010, p. 75).

Thirdly, most of the adopted and sanctioned pedagogical approaches from the West are not appropriate for the centralized HES in Vietnam, which traditionally employs 'top down' teaching and learning approaches. When teachers and students have the least power in the system, it might be expected that they would become increasingly passive in their teaching and learning. In Vietnam, when university teachers have been expected to employ new teaching strategies, they have not tended to argue against them, but little evidence exists regarding their ownership of these ideas or interest in implementing them, beyond compliance. The lack of adequate support for the teachers who have been educated in a traditional teacher-centred system has also hindered the national effort to manage university teachers. Teaching in Vietnamese universities continues to be critiqued harshly as not encouraging deep learning through interactive modes (Harman & Nguyen Thi Ngoc Bich, 2010, p. 74). Meagre additional resources for teaching and learning also contribute to this resistance to the 'preferred' Western teaching strategies. In most courses, exams are also required to ensure that the teachers deliver the required knowledge (Hoang Tuy, 2009; Ngo Doan Dai, 2006; Stephen, Doughty, Gray, Hopcroft, & Silvera, 2006) thereby further demonstrating the systematized management Vietnamese academics.

Perhaps even more profoundly, however, is the impact of the low level of research capacity in Vietnamese higher education and the overriding focus on applied and practical learning, rather than the theory-oriented research (Harman & Nguyen Thi Bich Ngoc, 2010). The lack of research capacity and funding appears to have lead policymakers to rely on the foreign 'advisors' who do not necessarily understand the local context. Their advice to implement Western policy and educational practices, which are not readily appropriate for the socio-cultural and infrastructure situation in Vietnam, has not led to a thriving HES. The effect instead has been that the Vietnamese HES has become externally managed, and the aim of this nation state to reform its HES has not been adequately served by Western neocolonial advice.

CONCLUSION

Vietnamese universities are managed by not only the government authority (MOET), but of greater concern, by external Western forces. The self-reliance of local Vietnamese academics has been disregarded and the dependency on both the money and the expertise of the West has increased. The systematic management of higher education in Vietnam privileges the West into a superior and expert role. However, the Western educational approaches, though considered 'modern,' 'effective' and therefore 'better,' do not appear to have improved higher education in Vietnam. The process of policy copying rather than policy learning (Deem, Mok, & Lucas, 2008) may, in the longer term, do more harm than good for the development of the Vietnamese HES.

NOTES

[i] For details see http://vanban.moet.gov.vn/?page=1.15andscript=viewdocandview=15566andopt=brpage.
[ii] For further discussion of curriculum in higher education see Barnett and Coates (2005), Clegg (2011), Lee (2005), Rizvi and Lingard (2010), and White (2012b).
[iii] This issue is also raised in Burnard and White (2008) in relation to Western school level pedagogy.

REFERENCES

Albach, P.G. (1971). Education and neocolonialism. *Comparative Education Review, 15*(2), 237-239.
Altbach, P.G. (2004). Globalisation and the university: Myths and realities in an unequal world. *Tertiary Education and Management, 10*(1), 3-25.
Albach, P.G., & Kelly, G.P. (1978). *Education and colonialism*. New York: Longman.
Apple, M.W. (2005). Are markets in education democratic? Neoliberal globalism, vouchers, and the politics of choice. In M. Singh, J. Kenway, & M.W. Apple (Eds.), *Globalizing education: Policies, pedagogies, and politics* (pp. 209-230). New York: Peter Lang.
Ashcroft, B., Griffiths, G., & Tiffin, H. (1998). *Key concepts in post-colonial studies*. London: Routledge.
Barnett, R., & Coates, K. (2005). *Engaging the curriculum in higher education*. Maidenhead: Open University Press.
Blackmore, J. (2003). Tracking the nomadic life of the educational researcher: What future for feminist public intellectuals and the performative university? *The Australian Educational Researcher, 30*(3), 1-24.
Burnard, P., & White, J. (2008). Creativity and performativity: Counterpoints in British and Australian education. *British Educational Research Journal, 34*(5), 667-682.
Bray, M. (1993). Education and vestiges of colonialism: Self-determination, neocolonialism and dependency in the South Pacific. *Comparative Education, 29*(3), 333-348.
Castells, M. (2000). *The rise of the network society*, 2nd ed. Oxford: Blackwell.
Chilisa, B. (2005). Education research within postcolonial Africa: A critique of HIV/AIDS research in Bostwana. *International Journal of Qualitative Studies in Education, 18*(6), 659-684.
Clegg, S. (2011). Cultural capital and agency: Connecting critique and curriculum in higher education. *British Journal of Sociology of Education, 32*(1), 93-108.
Codd (2005). Teachers as 'managed professionals' in the global education industry: The New Zealand experience. *Educational Review, 57*(2), 193-206.
Davies, B., Browne, J., Gannon, S., Honan, E., & Somerville, M. (2005). Embodied women at work in neoliberal times and places. *Gender, Work and Organization, 12*(4), 343-362.
Deem, R., Mok, K.H., & Lucas, L. (2008). Transforming higher education in whose image? Exploring the concept of the 'world-class' university in Europe and Asia. *Higher Education Policy, 21*, 83-97.
Deem, R., & Brehony, K. (2005). Management as ideology: The case of 'new managerialism' in higher education. *Oxford Review of Education, 31*(2), 217-235.
Harman, G., & Nguyen Thi Bich Ngoc. (2010). The research roles of Vietnam's universities. In G. Harman, M. Hayden, & Pham Thanh Nghi (Eds.), *Reforming higher education in Vietnam* (pp. 87-102). London: Springer.
Harman, K., & Nguyen Thi Ngoc Bich. (2010). Reforming teaching and learning in Vietnam's higher education system. In G. Harman, M. Hayden, & Pham Thanh Nghi (Eds.), *Reforming higher education in Vietnam* (pp. 65-86). London: Springer.
Hayden, M., & Lam Quang Thiep. (2010). Vietnam's higher education system. In G. Harman, M. Hayden, & Pham Thanh Nghi (Eds.), *Reforming higher education in Vietnam* (pp. 15-30). London: Springer.

Hoang Tuy. (2009). Educational crisis: Reasons and solutions in globalization challenges. *Vietnamese Comparative Education, 2009*, 1-9.

Hofstede, G. (1983). National cultural in four dimensions. *International Studies of Management and Organization, XIII*(1-2), 46-74.

Lee, A. (2005) Knowing our business: The role of education in the university. Discussion paper. Retrieved from Australian Council of Deans of Education website: http://www.acde.edu.au/publications.html

Lyotard, J.F. (1984). *The postmodern condition: A report on knowledge* (La condition postmoderne: Rapport sur le savoir, 1979). (Trans G. Gennington; B. Massumi). Manchester. Manchester University Press.

Marginson, S., Murphy, P., & Peters, M.A. (2009). *Global creation: Space, connection and universities in the age of the knowledge economy*. New York: Peter Lang.

McWilliam, E. (2004). Changing the academic subject. *Studies in Higher Education, 29*(2), 151-163.

Ngo Doan Dai. (2006). Higher education in Vietnam. In M.N.N. Lee, & S. Healy (Eds.), *Higher education in South-East Asia* (pp. 219-250). Bangkok: The UNESCO Asia and Pacific Regional Bureau for Education.

Pham Lan Huong, & Fry, G.W. (2004). Education and economic, political, and social change in Vietnam. *Educational Research for Policy and Practice, 2004*(3), 199-222.

Pham Thanh Nghi. (2010). The Higher education agenda: a vision for 2020. In G. Harman, M. Hayden, & Pham Thanh Nghi (Eds.), *Reforming higher education in Vietnam: Challenges and proirities* (pp. 51-64). London: Springer.

Phuong-Mai Nguyen, Elliott, J.G., Terlouw, C., & Pilot, A. (2009). Neocolonialism in education: Cooperative learning in an Asian context. *Comparative Education, 45*(1), 109-130.

Phuong-Mai Nguyen, Terlouw, C., & Pilot, A. (2005). Cooperative learning vs Confucian heritage culture's collectivism: Confrontation to reveal some cultural conflicts and mismatch. *Asia Europe Journal, 3*(3), 403-419.

Rizvi, F. (2007) Postcolonialism and globalization in education. *Critical Studies – Critical Methodologies, 7*(3), 256-263.

Rizvi, F. (2004) Globalization and education after September 11. *Comparative Education, 40*(2), 157-171.

Rizvi, F., & Lingard, B. (2010). *Globalizing educational policy*. London: Routledge.

Rose, R., & Mackenzie, W.J.M. (1991). Comparing forms of comparative analysis. *Political Studies, 39*(3), 446-462.

Singh, M., Kenway, J., & Apple, M. (2005). Globalizing education: Perspectives from above and below. In M. Singh, J. Kenway, & M.W. Apple (Eds.), *Globalizing education: Policies, pedagogies, and politics* (pp. 113-134). New York: Peter Lang.

Sparkes, A.C. (2007). Embodiment, academics, and the audit culture: A story seeking consideration. *Qualitative Research, 7*(4), 521-550.

Stephen, W., Doughty, P., Gray, P.J., Hopcroft, J.E., & Silvera, I.F. (2006). *Observations on undergraduade education in computer science, electrical engineering, and physics at select universities in Vietnam*. Washington: Vietnam Education Foundation.

Strathern, M. (Ed.). (2000). *Audit cultures: Anthropological studies in accountability, ethics, and the academy*. Abingdon: Routledge.

The Vietnamese General Statistics Office. (2009). Higher Education Statistics. Retrieved from the Vietnamese General Statistics Office: http://www.gso.gov.vn/default.aspx?tabid=395andidmid=3andItemID=9053.

Tran Thi Tuyet. (2010a). Decentralization in higher education in Vietnam: Some practical issues. Paper presented at the Decentralization in Higher Education from a Global Perspective: Implication for Vietnam and the Region, Ho Chi Minh city, Vietnam.

Tran Thi Tuyet. (2010b). Neocolonialism in education: Student centred learning in the Vietnamese context. Paper presented at the Global Education Conference, the Global Education Collaborative.

Trompenaars, F., & Hampden-Turner, C. (1997). *Riding the waves of culture*. London: Nicholas Brealey.
Johnson, A.T., Hirt, J.B., & Hoba, P. (2011) Higher education, policy networks, and policy entrepreneurship in Africa: The case of the Association of African Universities. *Higher Education Policy, 24*(1), 85-102.
Watson, K. (1994). Technical and vocational education in developing countries: Western paradigms and comparative methodology. *Comparative Education, 30*(2), 85-97.
White, J. (2012a). Scholarly identity. In T. Fitzgerald, J. White, & H.M. Gunter (Eds.), *Hard labour? Academic work and the changing landscape of higher education* (pp. 42-67). Bingley UK: Emerald.
White, J. (2012b). Turning a scholarly blind eye. In T. Fitzgerald, J. White, & H.M. Gunter (Eds.), *Hard labour? Academic work and the changing landscape of higher education* (pp. 92-118). Bingley UK: Emerald.
Wickens, C.M., & Sandlin, K.A. (2007). Literacy for what? Literacy for whom? The politics of literacy education and neocolonialism in UNESCO and World Bank sponsored literacy programs. *Adult Education Quarterly, 57*(4), 275-292.
Young, R. (2001). *Postcolonialism: An historical introduction*. Malden, MA: Blackwell.

Thi Tuyet (June) Tran
La Trobe University, Melbourne, Australia; and
Vietnam National University, Hanoi

Julie White
La Trobe University, Melbourne, Australia

JULIE WHITE, JANE GRANT, JEAN RUMBOLD AND BRUCE RUMBOLD

PRESUMPTUOUS METHODOLOGY

Troubling success, failure and research design

PRELUDE

In Melbourne, the month of May signals the end of autumn and that the academic semester is in full swing. The mornings are chilly and it's difficult to get out of bed. At the other end of the day, the clocks have been wound back and it gets dark early. Like the lightness of summer, all the welcoming events for new students have disappeared and across the campus the first years are assumed to have settled in to the rhythm of university life. To-do lists in iPhones and on scraps of paper pinned to notice boards detail all the reading that needs to be done; the meeting times with other students to complete assignments; the lab report due dates and essay criteria. The gloss of being a new university student has dulled and the faults of new friends are becoming apparent. It's the season for catching colds, and for whatever other lurking health concerns to come to the fore. There are fewer parties now. The once distant exams no longer seem so far away, and most have turned their attention to their studies as the pressure to measure up, to perform and to produce intensifies.

Figure 1

M. Vicars et al. (eds.), Discourse, Power, and Resistance Down Under, 161–173.

Jean's evocative photographs delighted the whole research team and early in the project we spent time examining, prioritising and selecting different images. The photographs depict sites around our University: the tables in the agora, the long open air corridors, and the electronic glass doors, which readily identify La Trobe and establish (or so we thought) a connection between the students and 'their' space. It was much later, when there were no participants, that we returned to the photographs and thought more closely about the assumptions they contained. And in turning the lens inwards (Ellis & Bochner, 2000) we noticed how we had inadvertently predetermined the narratives we were expecting to hear from those young people – those who had registered with the Diversity and Equality Centre, and who had ticked the box on the enrolment form that said 'Yes, I have a disability.'

INTRODUCTION

This chapter reports on a small funded project that did not go to plan. In the first part of this chapter the design of the study is outlined and the research team is considered. This is followed by discussion of the ethics approval process in the context of global higher education reform and what this means for researchers. However, in troubling the processes and performances involved in gaining ethics approval in universities, and in determining what constitutes success and failure in research, we are also troubled by the presumptions of our methodology. The second part of this chapter draws on a recorded discussion with the researchers on the failings of the project. The main argument put forward here is that 'good' research requires more than the gathering of evidence and the implementation of a neat and HREC approved design. We argue that projects that go wrong contain valuable knowledge which is often overlooked as a consequence of narrow definitions of what constitutes successful research. Further, we argue that failure is, in fact, an opportunity for researchers to critique university processes and practices and to scrutinise and learn from their own.

THE PROJECT

The purpose of the Narratives of Transition in Education (NOTE) project was to investigate the experiences of students who registered with the Equality and Diversity Centre as having a disability. How students with additional challenges of disability experienced their first year in higher education was the focus, and we hoped to identify, through this project, what helped and hindered these students in making the transition into the university. Further, we hoped to do so in ways that would recognise their perspectives and their voices. In the process of designing the study and applying for funding, we decided to offer workshops and a website where students could reflect on their experiences and share them through narrative and arts-based forms. Staff at the Equality and Diversity Centre offered support for recruitment and were consistently positive and helpful.

THE RESEARCHERS

The research team promised a rich collaboration as each researcher brought a different disciplinary background to the project. Julie White, from the Faculty of Education focuses her research on identity and social justice and recently finished working on a large study funded by the Australian Research Council that investigated identity, friendship and schooling experiences of young people with chronic illness. Jean Rumbold's background in counselling psychology, participatory action research and arts based inquiry is reflected in her work in the School of Public Health and at MIECAT, an organisation focused on education and research in the tertiary sector related to experiential and creative arts. At the time of the study Bruce Rumbold was the coordinator of a core first year subject within the Health Sciences Faculty. He is published in three fields: health sociology, physics and theology and has PhDs in each of the second two disciplines. Jane Grant was employed to work on the Note project. She has a PhD in Literary Studies and a research interest in biography.

Thus each of us brought different perspectives to the project. We anticipated learning from each other in ways that would enrich the project as well as inform our individual research practices. Julie could draw on her recent experience in the longitudinal study that used narrative and visual approaches. Jean would draw upon her experience in participatory arts-based research, and could support any participants wanting to use photography. The team agreed that her counselling expertise would be helpful in this study as we anticipated participant vulnerability and realised that particular sensitivities would be required. Bruce would bring insights from his recent work in the common first year Health Sciences program as well as his wider interest in the transition to university honed by participating in the 2008 European First Year Experience Conference. His research in spirituality and his role as Director of the La Trobe University Palliative Care unit also indicated different research perspectives. And Jane brought her considerable skill and expertise in writing biography that would support students to shape their narrative texts. All four of us had children who were, or recently had been, undergraduate students, so we thought we were sufficiently attuned to issues of first year students.

ETHICS APPROVAL

With the timing of the project in mind, Jean and Julie began working on the application to the Human Research Ethics Committee (HREC) in November 2009, at the end of the Australian academic year, although we did not complete it until late January. The timing of the project was always going to be important. We submitted the HREC application on January 30, 2010. The detailed 21-page checklist anticipated all sorts of unlikely events and took us several days to prepare. It didn't really address any of the sensitive and complicated issues we have subsequently been forced to consider, but asked specific questions related to storage of data, number of participants, research aims and other issues related to the general issue of design and conduct of the research. Contemporary qualitative research is deeply complex and uncertain, and yet the documentary requirements HRECs impose have been informed by protocols and principles designed to protect the institution from risk (Cannella & Lincoln, 2007; Halse & Honey, 2007; Sikes & Piper, 2010). Derived from the Belmont Report (Department of Health, Education & Welfare, 1979) Australian HREC processes are based on ethical principals for biomedical and behavioural research, and the continuing relevance of this philosophical frame for contemporary research has been questioned elsewhere (White & Fitzgerald, 2010). Why all ethical issues in research are expected to be addressed before, rather than during or after the conclusion of projects, warrants further consideration. Arguably, ethical and moral considerations occur at every stage of research, but in the contemporary university HRECs take a compliance and checklist approach.

We received a memo from the HREC on April 14 requiring attention to three issues before approval for the study would be given. Our response was sent to the HREC on April 24 and final approval was granted for the project to proceed on May 3, 2010. A summary of the concerns raised by the HREC was:

1. The correct form was not used to explain the project to the participants and there was a minor concern about the wording used;
2. The HREC admonished the researchers for not proofreading the document adequately, for a number of typographical errors had been identified;
3. The researchers were required to include a statement in the 'Participant Consent Form' pointing out that additional permission needed to be sought to include in academic publications any images produced by participants;
4. Concern was expressed that there was potential for conflict of interest if any of the participants were also students of the researchers.

We dutifully made the changes required, as this is what researchers are required to do in the contemporary university. Like Halse and Honey (2010) we wanted to behave ethically 'in terms of complying with institutional ethics policy and being morally and ethically responsible' (p. 123).

The HREC requirements for this particular study did not ask us to think about the ethical and moral implications of the study beyond those that relate to design and potential conflict of interest – the latter predicated on a distrust of researchers' capacity to behave appropriately. However the HREC requirements had significant implications for the study. Firstly, the beginning of the Australian academic year is March, and by the time we received final approval for the study, it was too late to successfully access new students. As Sidani et al. (2010, p. 106) point out, 'The importance of recruitment is well recognized ... The timing and strategies for recruitment can influence potential participants' decision to enrol in a study.' Secondly, one of our researchers, Bruce Rumbold, coordinated a large first year subject in the Faculty of Health Sciences involving 1600 students from all health science disciplines, across all of the university's campuses. Because of perceived 'conflict of interest' we could not recruit through this subject, and effectively could not include any students from Health Sciences, the largest Faculty at the university. In hindsight it might have made sense for Bruce to withdraw from the project at this stage, but this option was never discussed.

To expand a little further on this point, problems with recruitment frequently occur due to the researchers' failure to liaise appropriately with the stakeholders who will be involved in recruitment. This was not the case with our primary stakeholder, the Equality and Diversity Unit, who were engaged and supportive. But the other stakeholders we needed, staff members who worked with student groups that included students who had registered a disability, were actually taken out of the equation by an ethics committee interpretation that characterised teaching relationships as inherently coercive. In hindsight, again, it would have been prudent to have challenged that ruling.

Finally, it can be argued that the methods employed in research should be consistent with the disciplines that inform that research. Here again the ethics process focused upon a linear, risk-reducing recruitment strategy that stood in marked contrast to the actual work we hoped to undertake with the students.

HIGHER EDUCATION CONTEXT

How universities 'work' has undergone significant change in recent times and the terms 'modernised' represents the new version where the external and internal 'hypersteering' (Brennan & Zipin, 2010) has lessened the autonomy of individual academics. This autonomy has been replaced by audit, quality, managerialism and performativity in a global higher education shift (Fitzgerald, White & Gunter, 2012; Marginson et al., 2009; Rizvi & Lingard, 2010). Few academics can escape being 'managed' (Blackmore, 2003) and this applies to research and ethics processes as well as what counts and what is valued in research practices. Australia has recently followed the lead of universities in England and New Zealand, where England's Research Assessment Exercise (RAE, now called the REF) and New Zealand's Performance Based Research Fund (PRBF) have had significant impact on most aspects of university work and culture. Australia's Excellence in Research Australia (ERA) measures the worth of universities, who in turn measure the worth of individual academics and their research, often in terms of metrics rather than contribution to knowledge. The mantra of 'publications out, grants in' has become increasingly loud and insistent. Ethical behaviour in research is 'addressed,' once the correct boxes have been ticked in the long HREC forms that focus on anticipation of technical problems. Once 'approval' has been gained, HRECs do not show interest in research projects beyond eliciting regular progress reports, unless there are issues with compliance or breaches of conditions.

WHILE ROME WAS BURNING

We met regularly throughout the months between January and May, developing recruitment flyers, working out timelines, outlining the workshops, and drawing up lists of the equipment we would purchase, depending on participant interest and choice. The meetings were also an opportunity to raise problems and obstacles. Our lively, free-ranging and forthright discussions were informed by our very different backgrounds and perspectives and, if they led inevitably to the odd territorial dispute, these exchanges were always stimulating and a great deal of fun. In all, we believed that we had developed good research questions that would guide the study and had thought through the design of the study. We agonised over the recruitment posters and the photographs we would use. We were confident that our methods, processes and timelines allowed for participants from regional and rural campuses of the university to be included, and had discussed the issues of power and the need for sensitivity, although we have needed to revisit this more recently. At no time did it occur to us that we may not successfully recruit, and we worked diligently on detail of the study design while waiting for HREC approval.

However, it was Jane who drew our attention to the likelihood that students with issues of anxiety or mental illness could be reluctant to 'go public' in a project such as ours. And, over time, it dawned on us that we were actually asking a great deal of the participants. Instead of anonymously reporting on their experience to one researcher during an interview, with all the privacy involved in that, we were asking them instead not only to join a group of other students, but also to be prepared to express their challenges and concerns publically in images or narratives. An additional layer of presumption, on the part of the researchers, was that not only would these potentially vulnerable participants be willing to participate in workshops with strangers, and to do so in a group setting, but they would also be willing to write about their lives, or visually represent their challenges. We assumed perhaps that representing their experience in art forms would be easier than articulating them explicitly – that presentational knowing would be an attractive alternative to propositional knowing (Heron & Reason, 1997).

SCRUTINY AND METHODOLOGIC PRESUMPTIONS

What follows is an extract from a conversation recorded in April 2011, in which, using 'scrutiny' as a method of inquiry, the research team turned its attention to its own presumptions and processes.

> JR: I think we were interested in trying to understand something of the experience of first year students. Our funding was from Equity and Access so we took that particular slant of students who had registered with Equity and Access and their experiences of first year.
>
> JW: And to get the perspective of the young people themselves which we took into arts-based activities to try and get that, but we really wanted to get their perspective not the system, not the policy.
>
> JR: And we went to arts-based methods – it's ironic really – because we wanted to get a richer account of their experiences, you know, rather than

coming out with some sort of preformed survey where you tick the boxes. But I think it is probably one of the reasons, one of the many reasons, why we asked too much of the students.

… I do realise that I was thinking with a kind of educator hat on rather than with a researcher hat and I've had no difficulty recruiting people to do this kind of arts-based stuff [in the past]. And again the people doing that, obviously they knew me and have listened to me. When I got to think about that I don't think I would have responded to [the NOTE project].

JW: But I have had so much faith in that poster. I thought it is so beautiful and I thought how is anyone going to resist? I thought they would think 'That's my experience out there. I'm going to sign up for this.'

JG: I wanted to ask you about the photographs because I think they are brilliant photographs, but I also think they make a statement about alienation, loneliness, depression; they are laden with assumptions which trouble me when I'm thinking about them now.

JR: That's right.

JW: And if you are having a good day and you are a first year student you say 'I don't want to go there!' I never thought that at the time though.

JR: And if you were there, you wouldn't want to …

JW: Stay there.

JR: And the action of meeting up with strangers, you know.

JG: It was an open ended project where you didn't know anything about the experiences beforehand, but the more we have been talking and the more we look back on the design, it was making a lot of assumptions wasn't it?

JR: Yes.

JG: What do you think are some of the other reasons behind the project's failure to secure participants?

JR: We have covered timing;

JW: Insensitivity;

JR: Well the business of identifying as disabled;

JW: The recruitment; not being able to invite;

JR: Not being relational;

JW: And the poster flagging a whole lot of things.

JR: The Y Generation stuff: the fact that we were asking these people to commit to how many sessions ahead of time? I can't imagine them committing to coming to a party that far ahead!

JG: They were also first years.

JW: They are so unsure of themselves.

BR: The other thing I registered was the mental health aspect. Most of these kids are struggling. They are not going to take on other commitments when sometimes they can't even meet the deadlines the University imposes.

JG: In what ways do you think non-participation reflects the first year experience?

JW: We can only speculate, and they didn't sign up for a whole lot of reasons: one of them being that they were first years and it was all so much and all so overwhelming. I mean I have caught so many [of my own students] in tears …It's just all too big.

JR: Ironically we were trying to offer them a smaller place to belong, but of course they wouldn't have known that until they got there.

JW: We also had a whole lot of strings attached. We wanted them to bare their soul and tell us about things that they were not game to reveal possibly even to themselves. We wanted them to spill their guts and paint it, and photograph it … when you think about it, it was a bit of folly in a way.

BR: They didn't know this at that stage though did they?

JW: No.

JR: No they didn't … and it could have been as low key as taking a few snaps.

JG: So half the La Trobe students who register with a disability do so on the grounds of a mental health condition.[i] Do you think this might have any bearing on the issue of non-participation?

BR: Yes, judging by [a student I knew] with bipolar disorder. Tutorials were almost impossible for him because when he was manic he would be so over-engaged that people were scared of him, and when he wasn't he was seen as not participating. He couldn't cope with the tutorial system and he would never have been able to get through our re-designed curriculum, because it is based around enquiry and people are assigned to work in enquiry groups. 40 to 50 percent of their assessment is based around it and he would not be able to contribute appropriately to it.

JR: But even without people having a mental heath diagnosis…

BR: Introverts don't like it.

JR: Absolutely. So many people's learning style is not going to suit discussion based [learning].

JG: So what you are saying is that for people with complex mental health needs it is bad enough being in the tutorial system without signing up for a research group which was sort of tutorial based?

BR: Yes.

JG: Any thoughts on visibility, exposure and stigma in relation to the non-participation by the students?

JR: That's what we didn't find out about, whether they were experiencing any of those things.

JG: I meant in relation to their non-participation in this project; why they did not want to be involved.

JR: Well it would be consistent with Julie's research [from the Keeping Connected Project] … Students want to pass muster as ordinary and are not going to want to identify [themselves as having a disability]

JW: That was one of the earliest things to come out of the Keeping Connected Project.

JR: And if that was really strong in school age kids I doubt it is going to be very different in first year students.

JG: People with invisible disabilities find it harder to reveal them.

JW: The more obvious and the more external ... I think framing our discussion in terms equity rather than disability might take us further.

BR: That is the way medical thinking operates. You have normality and then you deal with deviance from the norm and the University runs on the notion of able-bodied participating students. Anything which is a divergence from that is treated as a deviation from the norm.

JW: There are people who say they know about the first year experience. There is the literature. There is a journal or two and conferences and all the rest of it, but first year experience using arts-based inquiry to explore that? There's bugger all. We have plenty of surveys. It is difficult. All universities should make allowances but what allowances do they make? What allowances should they make? We will never get there if we are not allowed to ask questions about these experience, which was what we aimed to do, which was 'Who are you? And what's it like being in your shoes at this point in time?'

Denzin and Lincoln's (2011, p. 15) point about the field of qualitative research being defined by 'a series of tensions, contradictions, and hesitations' neatly characterizes our study. Operating with the best of intentions, we wanted to hear about the transitional experiences of first year students with disabilities which we hoped would provide a greater understanding of the effectiveness of university accommodations. And yet we failed to register an inherent tension which should have been obvious both to us and to the Ethics Committee. As Julie and Jean noted, a key finding of Julie's research in the Keeping Connected study into school students with chronic illness was their resistance to being categorised as different from other so-called able students. And yet this was precisely what we were asking the first year students to do in the NOTE project. And in the La Trobe context, where over half of the students who register with a disability do so on the grounds of a mental health issue, visibility and the fear of stigmatisation could arguably be even more of a pressing concern.

As we explored in this discussion, another central and related flaw of our research design was our negative preconceptions about the experience of being a first year student with a disability. While our poster may have stated our aims as wanting to find out what helps and hinders the transition of first year students with disabilities (the good experiences as much as the bad), the tonal browns of our accompanying photograph with its focus on a distant shadowed figure seen through glass electronic doors is an unequivocal statement of loneliness. In other words, from the very start our invitation may have unintentionally excluded those for whom the transitional experience was positive and who may well have had a great deal to tell us about what worked for them.

CONCLUSION

As Wright Mills (1959, p. 139) exhorted, 'You must learn to use your life experience in your intellectual work: continually to examine and interpret it. In this sense craftsmanship is the centre of yourself and you are personally involved in every intellectual product upon which you may work.' We can only speculate as to why the students did not participate in our project: we can never know. And yet there is in fact a great deal that we can learn from this project. While evaluating the ethical processes of the University raises necessary questions about the limited involvement and narrow ethical concerns of universities, the need for researchers to keep questioning their own preconceptions and to continually 'examine and interpret' the interplay of their life experience and their intellectual work is perhaps the most salient lesson leant here.

The process reported upon here raises questions about accountability and power. The HREC ensures accountability by administering ethical assessment processes that focus principally upon managing attribution of fault and blame, should these questions arise. Ethical risks that might arise through relationships between researchers and participants are minimised by requiring that these be constrained to that of researcher and 'subject.' Researching pre-existing relationships, such as those with family, friends, colleagues, clients, students, requires detailed justification: for it is assumed that pre-existing relationship may bias selection and findings, and correspondingly that being a stranger will not. Nor does the HREC envisage situations in which the very process of research leads to 'real' human relationships: relational ethics are not part of the remit of the committee (Ellis, 2007). According to this view, research should not follow the normal pathways of learning or therapy, where relationship is integral to learning and change, and power is seen to have constructive possibilities that can be explored through negotiation. No-one would argue that accountability is unimportant, or that the coercive possibilities of power should not be confronted. But is possible to argue that formalised and stereotyped forms of compliance may in themselves be limiting and coercive (O'Reilly et al., 2008).

These issues come to a head in research that endeavours to be reflexive. Compliance with HREC-style ethical assessment processes becomes a primary concern of the researchers making application, focusing their attention on possible risk more than upon constructive possibility. Further, the reliance upon an exchange of documents works against negotiation between researchers and the HREC. To appeal a condition means delaying implementation by weeks; it is easier to accept further limitations in order to be able to do some research rather than none at all. Clearly, in this case, we as researchers should have been able to look beyond our concerns with compliance and be more aware of what we were attempting as well as what compliance might cost in terms of the success of the project. We should have been able to resist and transcend the approach into which the ethics review process directed us: but we failed to do so. What we now see more clearly is that reflexive research would benefit from more than gatekeeping.

Ideally, the HREC would be competent to supervise the research process and consult to the ethical issues that arise as the project is implemented.

The HREC's unwillingness to engage with relational ethics seems particularly unfortunate in the light of data that indicate that it is distance and indifference, not mutual engagement as a fellow human being, that precipitates harm in therapeutic relationships at least (Francis, 2010, for example). We believe that it is time to revisit both the conceptualisation and practice of ethical review, shaping it in a way that is more appropriate to the participatory nature of the actual research process.

ACKNOWLEDGEMENT

This study was funded by a Higher Education Equity support Program (HEESP) grant at La Trobe University. We would like to thank Stephanie Chard, the Manager of the Equality and Diversity Centre at La Trobe University, and her staff, for their support, encouragement and positive regard, even when the project did not proceed along anticipated lines.

NOTES

[i] Data provided by Equity and Access show that 232 of the 542 students who registered as having a disability in 2010 did so on the grounds of a mental health condition.

REFERENCES

Blackmore, J. (2003). Tracking the nomadic life of the educational researcher: What future for feminist public intellectuals and the performative university? *The Australian Educational Researcher*, *30*(3), 1-24.

Brennan, M. (2010). Dividing the university: Perspectives from the middle. In J. Blackmore, M. Brennan, & L. Zippin (Eds.), *Re-positioning university governance and academic work* (pp. 115-132). Rotterdam: Sense.

Cannella, G.S., & Lincoln, Y.S. (2007). Predatory vs. dialogic ethics: Constructing an illusion of ethical practice as the core of research methods. *Qualitative Inquiry*, *13*, 315-335.

Cresswell, J.W. (2009). *Research design: Qualitative, quantitative, and mixed methods approaches.* Los Angeles: Sage.

Denzin, N.K., & Lincoln, Y.S. (Eds.). (2011). *The Sage handbook of qualitative research.* Los Angeles, Sage.

DePalma, R. (2010). Socially just research for social justice: Negotiating consent and safety in a participatory action research project. *International Journal of Research & Method in Education*, *33*, 215-227.

Department of Health, Education & Welfare. (1979). Belmont Report: Ethical principles and guidelines for the protection of human subjects of research. Report of the National Commission for the Protection of Human Subjects of Biomedical and Behavioral Research. Federal Register, 44, 23192-7.

Ellis, C. (2007). Telling secrets, revealing lives: Relational ethics in research with intimate others. *Qualitative Inquiry*, *13*, 3-29.

Ellis, C., & Bochner, A. (2000). Autoethnography, personal narrative, reflexivity: Researcher as subject. In N.K. Denzin, & Y.S. Lincoln (Eds.), *Handbook of qualitative research*, 2nd ed. (pp. 733-768). Thousand Oaks, CA: Sage.

Fitzgerald, T., White, J., & Gunter, H.M. (2012). *Hard labour? Academic work and the changing landscape of higher education.* Bingley, UK: Emerald.

Francis, R. (2010). *Independent inquiry into care provided by the Mid Staffordshire NHS Foundation Trust January 2005–March 2009*, Vols. I & II. London: The Stationery Office.

Halse, C., & Honey, A. (2007). Rethinking ethics review as institutional discourse. *Qualitative Inquiry, 13*, 336-352.

Halse, C., & Honey, A. (2010). Unraveling ethics: Illuminating the moral dilemmas of research ethics. In W. Luttrell (Ed.), *Qualitative educational research: Readings in reflexive methodology and transformative practice* (pp. 123-138). New York: Routledge.

Heron, J., & Reason, P. (1997). A participatory inquiry paradigm. *Qualitative Inquiry, 3*, 274-294.

Marginson, S., Murphy, P., & Peters, M.A. (2009). *Global creation: Space, connection and universities in the age of the knowledge economy.* New York: Peter Lang.

O'Reilly, M., Dixon-Woods, M., Angell, E., Ashcroft, R., & Bryman, A. (2009). Doing accountability: A discourse analysis of research ethics committee letters. *Sociology of Health & Illness, 31*, 246-261.

Rizvi, F., & Lingard, B. (2010). *Globalizing educational policy.* London: Routledge.

Sidani, S., Streiner, D.L., & LeClerc, C.M. (2010). Mitigating the impact of external forces. In D.L. Streiner, & S. Sidani (Eds.), *When research goes off the rails: Why it happens and what you can do about it* (pp. 106-112). New York: The Guilford Press.

Sikes, P., & Piper, H. (2010). Ethical research, academic freedom and the role of ethics committees and review procedures in educational research. *International Journal of Research & Method in Education, 33*, 205-213.

White, J., & Fitzgerald, T. (2010). Researcher tales and research ethics: The spaces in which we find ourselves. *International Journal of Research & Method in Education, 33*, 273-285.

Wright Mills, C. (1959/2010). On intellectual craftsmanship. In W. Luttrell (Ed.) (2010), *Qualitative educational research: Readings in reflexive methodology and transformative practice* (pp. 139-145). New York: Routledge.

Yates, L. (2004). *What does good educational research look like?: Situating a field and its practices.* Maidenhead, UK: Open University Press.

Julie White
Faculty of Education
La Trobe University, Melbourne, Australia

Jane Grant
Faculty of Education
La Trobe University, Melbourne, Australia

Jean Rumbold
School of Public Health & Human Biosciences
La Trobe University, Melbourne, Australia

Bruce Rumbold
School of Public Health & Human Biosciences
La Trobe University, Melbourne, Australia

JULIE ARNOLD

THE ROLE OF ARTFUL PRACTICE AS RESEARCH TO *TRACE* PRESERVICE TEACHER EPIPHANIES

Figure 1. Participant M's art work

INTRODUCTION

This chapter discusses the appropriateness of art-based research methods in deliberating on pre-service teacher education. Based on a combination of interviews, filming and painting, participants are asked to reflect on learning and values which will provide strong foundations that will underpin their future teaching practice. The process is complex and demands the attention of the researcher to examine the visual artefact, text, movement and voice of the pre-service teacher. The chapter explains the quest for ethnographic research data through multiplicity of combinations with an emphasis on the co-created artform. In looking for new ways to make meaning this work exposes written, spoken and the painted forms, to make possible new research understanding and knowledge.

M. Vicars et al. (eds.), Discourse, Power, and Resistance Down Under, 175–187.

BACKGROUND

While the content of interviews is important, it is the context of the interview that often assists the researcher in procuring richness. Providing a structure for the interviewees to speak freely is particularly important to the quality of data. In research terms this may be referred to as 'social' or 'political' power where typically research demands equity from all involved when interviewing. A safe and equitable space is required, but it is the responsibilty of the researcher to provide a elegantly crafted platform that will create the beauty, compassion and the finest senstivity to untap and assist to devolve the real and often elusive resonating stories. Seeking the elusive story is particularly important and I do this with art at the core of the research process. As a result this chapter outlines an experimental methodology that challenges traditional ways to find meaning that is worthy for the qualitative researcher and the process can be rigorous.

The research outlined within this chapter uses art making to relinquish rich stories and mindshift from pre-service teachers and trace their personalised professional learning. The power and relevance of 'mindshift' in terms of shifting perspectives in pre-service teacher education seems to be important to the role in the development, socialization, and induction of teachers to the world of holistic practice (Cochran-Smith & Lytle, 1999).

By highlighting the artful methodological process, the creating of visual art images acts as a platform and space for the participant that not only supports and expore their stories, but signal accounts and experiences from their life learning. Additionally, in this research project the reflective conversations were filmed and subsequently transcribed by the researcher to trace the significant moments of learning that emerge around pre-service teacher learning and identity. Using stories and art work in this instance, is important to the reflection process for tracing mindshift during teacher development. Whilst this chapter has specific content relating to the work of teaching the researcher proposes that the language of art is beyond a 1000 words. This method of inquiry can and should be used for visual learners especially. When reflecting on their practicum pre-service teachers may come to consider art making as an increasingly important responsibility and to see themselves as learners (Groundwater-Smith et al., 2003).

ACCESSING REFLECTIVE CONVERSATIONS

The research accessed reflective conversations from and with fourth year pre-service teachers, with the aim of describing the layers of learning experiences, critical moments and development involved in their pre-service teacher identity formation. In particular, articulating the reflective pedagogical paradigm that is inherent in the process of becoming not merely a teacher but a reflective practitioner. This reflectivity is at the core of the art making and the stories told whilst making art.

Artful practice is used to encourage conversation and is often implemented for story telling particularly in indigenous and feminist socio-cultural groups (Martin & Booth, 2006). Stories and images created on canvas using, ink, acrylic and oils are intentional and representative of the reflective educational journey. Additionally, the act of telling and painting concurrently the narratives emerge as a method of reflecting and termed meta-cognitive landscapes.

The meta-cognitive landscapes assisted by the reflective conversations are analysed. Alvesson & Sköldberg suggest, 'Conversations are analysed by breaking them up into segments (in analogy with linguistics). It is then found that certain types of segment or "slots" appear, irrespective of the content of the conversation' (Alvesson & Sköldberg, 2000, p. 42). Here, conversation analysis, although not the focus of the research is equally important as the art making. 'Conversation analysis' (Atkinson & Heritage, 1984; Boden & Zimmerman, 1991; Sacks, 1992) assists in painting the whole picture. Heritage (1984) in Holstein and Gubrium (1994, p. 226) suggests, conversation analysis has three basic prerequisites:

1. There are recurring structural features in ordinary conversation, irrespective of psychological characteristics of the participants.
2. Conversation is context-bound, such that it is both productive of and reflects the circumstances of its production.
3. These two properties characterise all interactions, so that no detail can be dismissed as out of order, accidental or irrelevant to the ongoing interaction.

As data, the stories and images are not selected *ad hoc*; they are selected from the systematic process of assisting in the broad experiences based on the day-to-day experiences and will provide evidence of what they consider crucial in their own learning. Following responses to the semi-structured interview[i] questions, the interview commences however, the communicative nature of the process tends to lapse into conversation rather than adhering to a normative questioning technique. Early conversation stems from the ideas derived from questions as a mode to eventually reach a broader and rather substantial end product.

Original transcripts, film and artwork intertwine to ensure an authentic and trustworthy testimony to the voice of participating pre-service teachers. The authentic liguistical tones of the pre-service teacher although varied, are recognised characteristics and subsequently valued. At times, during the interview the voice takes on a cultural and compassionate tone or agency. This is not always apparent in the transcript and often there is opportunity to describe fully the context, motivation and other aspects described in the methodology which is not only narrative and voiced through conversation, but expressed through voice quality, gesture and image.

RESEARCH CONTEXT

Focusing on the existing pre-service teacher education practices in the light of socially reflective view in the existing context (Foucault, 1988), this research relates to thoughtful post structural approaches and pedagogies accepted by contemporary theory. Additionally, the author as artist-researcher holds that the reflective approach acts as a framework for developing deeper insights into current thinking and theory-practice debates in the wider research community. The 'reflective' approach means that attention is paid to the interpretive, political and rhetorical nature of empirical research (Alvesson & Sköldberg, 2000).

> Reflection means interpreting one's own interpretations, looking at one's own perspectives from other perspectives, and turning a self critical eye onto one's own authority as interpreter and author. (Alvesson & Sköldberg, 2000, p. vii)

The research in this instance draws on awareness to a broad range of insights into epiphanies, experience and interpretive acts, into political ideological and ethical issues of the social sciences, and connected to the voice of the participant.

Whilst research undertaken by Cochran-Smith and Lytle (1999) focuses on the power and relevance of mindset in pre-service teacher education and this is seen to be important to the role of development, socialization, and induction of teachers to the world of holistic practice (Cochran-Smith & Lytle, 1999) this work is substantially different. Using art and conversation it is a *tracing* of 'when' and 'how' the transition of mind-shift in teacher learning takes place and builds upon contemporary knowledge to assist the development of pre-service teachers.

CONTRIBUTION TO KNOWLEDGE

Successful educators are those who are able to be critical of their own limitations and who, in the process, create a vision of pathways that lead to an enriched professional status (Schulman, 1992). The process is cyclic and continuous. The process of self-discovery in this instance, generates a particular kind of 'mindshift' that is likely to influence their professional philosophy and practice.

The author contends that teachers in training become aware of their multidimensionality, which implies a willingness to conceptualise their evolving self as teacher. Such a re-contextualisation requires a willing disposition, and an aptitude to adopting varied teaching capacities and circumstances. Reflecting on practicum has become one way for the pre-service teachers to consider an increasingly important responsibility to see themselves as learners (Groundwater-Smith et al., 2003).

Artful practice as research, is to further understand experiences and development of pre-service teachers as they move to professional and holistic practitioners It asks how pre-service teachers know that they are informed and functioning in a 'rich' manner.[ii] The broader focus of this qualitative ethnographic form of interpretative inquiry has generated new thinking leading to an understanding about artful practice within the broader research arena.

Who are the participants and how can we trace their thinking?

The research project methodology examines the layers in the development of 10 pre-service teachers, and in particular how they attend to emancipatory action

(Friere, 1996) and self knowledge (Giroux, 1997). The researcher believes the themes are assumed to be around the following:

1. self-efficacy;
2. conceptualisation and actualisation;
3. identity;
4. cultural /structural/ historical competence;
5. critical awareness;
6. independent inquiry/inquiry approach;
7. social capital and
8. lifelong learning.

An introduction for each participant prior to interview routinely begins by the researcher outlining the activity, materials available and the time frame. A comfortable environment is foremost and once achieved, art making and conversations begin. To clarify, the researcher routinely begins by discussing the term 'mind shift' with each pre-service teacher participant. The interview below for example, is typical of the approach for each participant. The inclusion of a number 3-4 Semi-structured Interview (SSI) questions enabled a starting point for each participant involved in the research.[iii]

PARTICIPANT M

M is male and from Arabic background. He is mid 20's and comes from the western suburbs of Melbourne. He has experience in nursing before deciding on teaching as a career. Parts of his interview below identify major themes and links to the artful practice as a conduit for rich conversation.

R[iv]: The topic we are discussing is the notion of mind shift in terms of becoming a teacher. So, what we are going to focus on is ... What were the trigger's that you can identify that has sent you on the journey to teaching? But first of all what do you understand by mind shift'?

M[v]: I think the word itself .... You know the mind shifting and its influence from one state to another.

M begins to roll out the ink on the canvas. The line he makes has a life of its own and takes its own pathway slowly and sometimes quickly as M lifts the canvas at times to create the unchartered marks the ink leaves. He then points to the line to highlight his next statement.

R: Some theorists have used the word 'mind shift,' 'enlightenment,' 'epiphany,' 'threshold concepts' and all sorts of terms to describe mind shift.

M: (Pointing to his initial lines within the painting). It's like you're on a long journey and it goes onto different paths. But look honestly, I believe that the

> reason that I guess this journey that I've gone on, is because of the teachers' themselves ... it's because my teachers were my inspiration and I believe that they were who I aspire to be and I honestly believe that. This society has made me who I am today. You know, and some of my friends have said 'You have opportunity and [now] you have the responsibility.' So with opportunity, comes responsibility, and I honestly believe that this is something that has helped me to get to where I am today. I've always wanted to do teaching; I've just never had the opportunity to get into it any earlier.

EMERGENT THEMES

There are a number of themes arise throughout the discussion. I would now like to examine the articulated responses from pre-service teacher M in relation to a series of significant themes. From the discussion above two themes emerge and following each commentary emergent themes that contribute toward professional mindshift are noted:

- Theme 1: Opportunity and access
- Theme 2: Journey

The context of journey can range from a trip or voyage to a predetermined or unfamiliar destination. By examining M's articulated and visual description – 'long journey,' he clearly defined the concept as a fluid and somewhat unchartered. Simultaneously, he articulates and visually describes the term. Additionally, he contextualises even more by demonstrating the 'long journey' moves and winds in different directions so paths are reinforced by the kinaesthetic response.

The interview is continued whereby, questions are raised as in naturalised form of conversation.

> R: So why teaching? What was the pivotal force or driver?

> M: I finished my degree in nursing, you know. Honestly, I think it's because I just knew that nursing would not have been for me. I couldn't, I just don't think I could've actually done it.

> R: What changed your mind?

> M: I think you know the reality of nursing, of ... you know ... going into a hospital and [seeing] people dying and I just couldn't cope with that. And the whole concept of a nursing home, it's not something that I have in my culture, you know. I am from an Arabic background, so we don't have nursing homes. And to see these people just lumped in there together was very shocking and I just couldn't see [why] that [was] happening and so I didn't want to do it.

Here, M provides are clear understandings and statements of important learning. The following themes become more relevant:

- Theme 3 Cultural Values
- Theme 4 Integrity and Action

Career change for M is guided by learning and is a trigger to tracing the values and action. Contextual issues support the shift and governed by integrity of M within this story. Interestingly, when M is thinking carefully about connections with hospital and dying, he moves from to the canvas and hesitates, and as he slows, his sentence formation also hesitates and he moves from the canvas and pauses, before approaching again to continue with an ink wash over the centre of the canvas.

Once again M engages in further conversation below and the themes from the commentary are identified.

> M: You know, they're human beings [elderly patients] and they also have their needs and they actually have their, (pause) … they have their (A little frustrated. He picks up spray and begins shaking the can vigorously). You have to respect them for who they are and at the same time you know (asking) … 'how would you like to be treated yourself'? (M makes eye contact as if appealing to the researcher). And that's something that I want to want to bring into education … I think I relate to children … because I am the oldest in the family and that's something that I've always had to do – help and assist members of the family.

M walks to collect a spray paint container and begins shaking it rigorously. He begins to paint while talking and sprays at the top of the canvas. He smiles and relaxes again.

> R: Ah, so what was the moment of epiphany? How did you know teaching was for you? What triggered it?

M looks at his hands and notices he has paint on each hand.

> M: I am just trying to think, what would it have been? Ummm. Yes, there is, well, one thing is … actually …

M has stopped painting and spread his arms slightly towards the R as if appealing through the following sentences and inferring these are important points.

> M: When I grew up you know … my teachers were everything … because there was a first language barrier between my parents and I, and the school. So, it was very hard for me to get along with my parents on that same level as I could with my teachers. But, my teachers were always there for me and so when I went to school it was who I was. At same time … Uhhhm, I feel that … that to be able to come here to university and to be a teacher is to relate … as a student, and at the same time it's something that I find enjoyable.
>
> I'm going to be really weird here, but I'm actually going to paint my fingers. (He begins to cover his fingers with a brush dripped in red paint.)

> A principal was asked why the parents didn't become involved in his school. He stopped them and said, 'NO' it's not that they don't want to it's just that they don't know how to go about it. So, that's something I believe affected me (printing his hand on the canvas). My parents came from overseas and they came during war and they had no formal education and when they came here they worked in you know ... in factories (now he uses yellow over the red paint still remaining on his hand). But you know, they really wanted education for us, but they honestly did not know, for example how to go about helping us any further. They would say 'Look, you know we can try to support you financially, but we don't know it's just a basic concept (education), but we don't actually know how to help you with that.' (He stamps a hand print on the canvas again this time more firmly.) That's why I keep saying you know, my teachers were my inspiration, because I turned to them because that's what teachers are there for.

Theme 5: Cultural and social capital

M's life and cultural experience inform and develop his personal and collective cultural values. Action and integrity along with passion and social justice are critical to his identity and are positive experiences that brought M back to university and indeed to teacher education. The strength of conviction is emulated by hand prints pressed firmly on the canvas. He presses with the weight of his body over the centre of the print. There may be a correlation in his response in choosing to use this technique to attempt to connect in a palpable way with this part of the conversation.

The following commentary reviews the concept of mindshift and the transformative positioning of M from pre-service teacher to teacher.

> R: Can you think of any moment that's a critical moment, where it's changed the way you think about becoming a teacher?

> M: I think that the pivotal moment probably was when I was in the classroom during my practicum. When we were there in schools I think that was a pivotal moment to say, 'OK.' 'Now you are realising you are becoming a teacher.' So, yes, it wouldn't have been something that I would have automatically have realised myself. When someone says something or you identify subconsciously, but you can't actually explain it to yourself until that moment – until the moment you say, "OK! I think that makes me a teacher."

Theme 6: Positioning of power / consciousness in the classroom is central to identity

During this conversation M picks up a large bottle of red ink. He unscrews the cap and pours it on to the canvas. Still speaking and without knowing where the ink may flow, he holds the two sides of the canvas and begins to take control of the ink. He rolls it about the canvas freely. He appears to be relaxed with the forward and backward flow of the ink. As with the conversation between him and the researcher, he is calm and cognitive of the free-flow of conversation and topic.

> R: What is it to think like a teacher?
>
> M: Being conscious of the people around you and the world. And I think you [teachers] have a responsibility. I honestly believe a teacher is someone who has the morals and the judgements of someone who's been given a gift.

Theme 7: Ethical considerations and stance.

This is a powerful statement and takes the concept of what is it to be a teacher to new levels of critique. Highly conscious of ethical codes, responsibility of morals and judgement are evident. M moves closer to the canvas. His head is low and he is attending to fine details in the painting. He has made some strong statements and is particularly attentive to small details in the art work.

> M: During a teaching session when planning lessons for students with other pre-service teachers, it took ten minutes (talking with others) to actually to understand everything I needed to do. It took ten minutes for me to say, 'OK, what I was doing was incorrect.' Well, team teaching is actually about gaining other skills from others or group members and then being able to use that to apply into my further teaching and that's what I did when we did team teaching. I thought that was really pivotal to know to be able do that – to know that teachers aren't just independent little bodies (moving his arms to provide description) and they are connected together one way or another. I believe that teachers don't have all the ideas themselves and so you ... use that network ... it's a network, (asking others) what works well and what doesn't work – 'Can you help me with this and what do you think I can do in relation to that?' Teachers are open to new things. They should be open to new things and be able to adapt; we're supposed to be able to that. That's what makes teachers stand out from other professions – to be able to do that. I think they see society in its most (slowing at this point, he makes eye contact) raw form because schools are the first point in what makes a child who they are in the future. At the same time, follow the changes that are happening in society, and they make it relevant to the student and consequently the rest of the people who live in that society.

Theme 8: Knowledge

Here, M establishes that teachers are not the 'holders of knowledge.' Teachers are collaborators and work together to connect knowledge. M recognises the importance of feedback to improve professional practice and subsequently, understanding that teachers are learners, rather than the deliverer of knowledge. During this next account, M moves from speaking directly to the researcher to reaching for a small stick and scraping on small amounts of black paint to make small deliberate markings.

> R: You were talking about playing teachers with your brothers and sisters. What happened?
>
> M: In our bedroom ... because my parents ... well they're smokers. They would sit around in the lounge room watching pre-recorded Arabic TV shows and Spanish shows and we did not want to sit in front of them. So, we would sit in our rooms and play teachers. We would get a book and we would mark the roll and we would say "We are doing maths today." It was just interesting to be able to do that – it was an interest to learn new things, I guess.

Theme 10: [Early] developmental play

This theme is is very relevant to reflect on and relates to the imporance of early learning through play.

> R: Can you explain a situation or describe a moment where you thought 'Ah ha!' 'I am the teacher.'
>
> M: I think this happened the first time I was in control of the classroom. You can think of it as in the physical sense; but it is the actual practicality of it. Maybe it's when the student turns around and says, "Thank you." When ***they*** (the student) actually appreciate what you're trying to do for them. So, I think that might be something that tells pre-service teachers that they **are** teachers; that they could be teachers and very good teachers. I think that it could be as simple as that.

Theme 11: Authentic praxis

This theme indicates the value of preservice teacher/student 'real life' relationships and being on the ground has significant depth to the understanding M has experienced.

> M: (Reaches for the black paint across the table). There are dark moments – they do happen at different times. So during the journey they do come along and do affect the way that you do things. Sometimes dark moments lead to lightness and I think that's why I have put in the yellow (pointing to art work). What do I do with a student who's not listening or not co-operative. I don't have the power to help them? So, I think that's one of the dark moments and at other times. It's the fact that you fail in your role as a teacher, I think. As an example, I was teaching a class and my mentor teacher

was there and I completely just shut down. I did everything that I shouldn't have done. He was an English teacher. It was really hard and I was scared of him. He was very critical of the way I undertook the lesson. But it was because he was critical that I think I learnt a lot. So, that was something I was able to use to my advantage because at the end of the day, I'm there to learn and basically, to satisfy the request that the VIT[vi] that has the authority to enable me to be a teacher. So I used that experience as a learning curve for me.

Theme 12: Assessment, reflection and feedback as learning

When the interview and the art work complete, M carefully observes both pieces of art work. He points to the researcher's painting (see below) and begins to interpret it from his perspective. He volunteers some feedback and critique of the researcher's art and in a closing comment, M states:

> M: Your art work [referring to researcher's art] looks like one of those landscape paintings. You have, the mountains and there ... that could be the shoreline here (pointing to the lower section of the canvas) and that could be the beach ... or something ... or it could be a lake and here (pointing again to the picture) ... it's like this journey that you take. Education is a journey. In fact, being a teacher or an educator is a constant journey.

CONCLUSION

If reflections are made visible through opportunities to narrate knowledge and holistic growth of the pre-service teacher then mindshift becomes clear for the emerging teacher. We can then begin to explore the mind shift and the process as an *epiphanic* phenomenon.

The process makes critical reflection possible. An example below illustrates this point. Here, in a commentary from M he articulates his mindshift on the subject of critical practice.

> M: If you can't critically reflect you won't be open to new ideas and you won't be able to, I think, progress as a competent teacher. That's critical – a teacher's not there just to teach. I think there's so much more to a teacher than meets the eye, because I don't think that there can be possible answers all the time. It just depends I guess on the situation and what you can do to change things, because teaching is not just something that is stagnant and just stays like that. It is actually something that evolves and teaching is always evolving so what worked yesterday does not work today.

Transcriptions of conversation act as data. Data provides themes that have emerged from deeply held experiences which are identified and recorded. Systematic analysis of the data enable identification of major themes and these have been noted. Additionally, images created in acrylic on canvas represent the reflective educational journey. The art making has the power to heighten the description and reflection beyond the text. The creation of a visual art image has enriched the recollections and the telling by providing space and imagined knowing.

Once the interview, conversation and art-making are complete, as researcher, I observe the space, reflect on the experience and ask, 'What have I missed? What data has not been revealed? Can I recall the important moments and what will I be able to give to the record?' I pack up the camera and label the film for reviewing and transcribing. The room is cleaned and things put back to resemble a classroom once again – and what of the space? The two paintings (Figures 1 and 2) are left to dry as testament to the story, themes and the encounter.

Figure 2. Researcher's art work in response to working with M

NOTES

[i] Three questions are asked at the introduction of each interview to begin the conversations.
[ii] Rich in this instance refers to productive, engaging and innovative professional pedagogical processes.
[iii] A written outline in the form of an invitation preceded a Plain English outline for each participant and was a starting point for informal discussion before SSI questions.
[iv] Researcher.
[v] Participant.
[vi] Victorian Institute of Teaching.

REFERENCES

Atkinson, J., & Heritage, J. (Eds.). (1984). *Structures of social action: Studies in conversation analysis.* Cambridge: Cambridge University Press.

Alvesson, M., & Sköldberg, K. (2000). *Reflexive methodology.* New York: Sage Publications.

Boden, D., & Zimmerman, D.H. (1991). Transcription appendix. In D. Boden, & D.H. Zimmerman (Eds.), *Talk and social structure* (pp. 278-282). Cambridge: Polity.

Cochran-Smith, M., & Lytle, S.L. (1999). Relationships of knowledge and practice: Teacher learning in communities. *Review of Research in Education, 24,* 249-305.

Danaher, G., Schirato, T., & Webb, J. (2000). *Understanding Foucault.* NSW, Australia: Allen & Unwin.

Denzin, N., & Lincoln, Y. (2000). *Handbook of qualitative research,* 2nd ed. London, UK: Sage Publications.

Dewey, J. (1963). *Experience & education.* New York, USA: Collier Books published by arrangement with The Kappa Delta Pi Lecture Series.

Freire, P. (1996). *Pedagogy of the oppressed,* 3rd ed. London, UK: Penguin Books.

Friere, P., & Macedo, D. (1987). *Literacy: Reading the word and the world.* London, UK: Routledge & Kegan Paul.

Foucault, M. (1980). *Power/knowledge: Selected interviews and other writings 1972-1979.* New York.

Foucault, M. (1988). The ethic of care for the self as a practice of freedom. In M. Foucault, *The final Foucault,* edited by J. Bernauer & G. Rasmussen (pp. 1-20). Cambridge: MIT Press.

Giroux, H. (1997). *Pedagogy and the politics of hope: Theory, culture and schooling.* Boulder, CO: Westview.

Groundwater-Smith, S., Ewing, R., & LeCornu, R. (2003). *Teaching: Challenges and dilemmas,* 2nd ed. Melbourne: Nelson Australia P/L.

Holstein, J.A., & Gubrium, J.F. (1994). Phenomenology, ethnography and interpretive practice. In N.K. Denzin, & Y.S. Lincoln (Eds.), *Handbook of Qualitative Research* (pp. 262-272). Thousand Oaks, CA: Sage.

Holstein, J.A., & Gubrium, J.F. (2000). *The self we live by: Narrative identity in a postmodern world.* New York: Oxford University Press.

Martin, E., & Booth, J. (Eds.). (2006). *Art-based research: A proper thesis?* Melbourne, Australia: Common Ground Publishing in association with Victoria University.

Sacks, H. (1992). *Lectures on conversation, Volumes 1 and 11,* G. Jefferson (Ed.) with an introduction by E.A. Schegloff. Oxford, UK: Blackwell.

Shulman, J. (1992). *Case methods in teacher education.* New York: Teachers' College Colombia University.

Van Manen, M. (1995). The epistemology of reflective practice. *Teachers and teaching: Theory and Practice, 1*(1), 33-49.

Vygotsky, L.S. (1962). *Thought and language,* edited by E. Hanfmann, & G. Vakar. Cambridge, MA: MIT Press.

Julie Arnold
Victoria University
Melbourne, Australia

MARY-ROSE MCLAREN

THE 'CONSCIENTIOUS SCHOLAR'

Balancing the roles of researcher, playwright and daughter

Let me begin with an historical account.

> It is 1941, and against advice, Churchill decides that the allies will enter the Balkans and support the Greeks against a German invasion. From the very beginning logic, geography, strategy and history determine that this is a battle that cannot be won (Beevor; Great Risk; Robinson). However, Menzies, prompted by Churchill, felt strongly that the allies must not leave Greece 'in the lurch' (Great risk, p. 9). So in the allied soldiers went, including the men of the sixth division of the AIF. (Swain, 2001)
>
> At first they had a lovely time (Morgan 8/10/41; Great Risk). And then the Nazi invasion began. Over a fortnight the allies were driven further and further south until they were evacuated off the Peloponnese. Some of these men, thinking themselves lucky to escape the Balkans, embarked on the Costa Rica, which was then bombed and sank (AWM/P05094.001).
>
> They scrambled as best they could onto other ships. Overloaded, and under fire, they were disembarked not in Alexandria, their planned location, but on the northern shores of Crete. No shoes, no supplies, no weaponry, they reformed as best they could, began daily marches and enjoyed the sunshine, scenery and hospitality. (Frean 1949; Morgan 8/10/41)
>
> And then, once more, the Germans came. Again, their position was untenable. Again, the only option was to flee, and so they trekked south, over the mountains, in the hope of being picked up at the southern port of Sfakia (Great Risk; Beevor; Robinson; Frean). They travelled by night, in pitch dark, and slept and hid in the bushes by day (Robinson; Morgan 8/10/41). Those who did not get to port, thousands of them, were taken prisoner (Great Risk).

Now, I would like to tell you a story.

> Fast forward to 2010 and my father's half-brother, born well after the war and a quarter of a century younger than my father, writes and asks: does my father have any information on the sinking of the Costa Rica? His Dad used to talk about it, and now my uncle would like to discover more. Did he ever say anything to my father about it?

M. Vicars et al. (eds.), Discourse, Power, and Resistance Down Under, 189–198.

> My father says, 'so I sent him a photocopy of the letter.'
>
> 'What letter?' I ask.
>
> 'The letter Dad sent to me.'
>
> 'When?'
>
> 'At the time.'
>
> And he produced a tin, and in the tin were eighty letters, all but one written during the six years my grandfather was at war, the six years when my father grew from a boy of ten to a young man of sixteen, a tin of letters that smell and feel of a different world.

My interest lies in the space between the history and the story – the space both inhabit but neither completely claims. In this case, the bridge between the history and the story is a box of letters. They have travelled physically across space and over time, from the theatres of war to home, from the 1940s to 2011. I am interested in the intellectual and emotional spaces in which these letters travel, and they ways that we view them as both personal and as public data. The story I have just told seems to be about my grandfather's experiences in World War II, but it is not. It is really about the way those experiences have rippled through two generations, and have impacted on identity, understanding and perceptions of family. The true story is documented in the letters but told in our daily lives.

As soon as I knew those letters existed I wanted to write a play. Why? A play is a form of writing that is very immediate – it shows peoples' experience rather than simply telling about them – it invites the audience to share those experiences. The actors' feelings are immediately visible, their thinking is present before the audience, and the elements of stagecraft assist in communicating feelings, ideas, and concepts to the audience without lots of words (Jackson, 2008). Plays give us the opportunity to find meaning, process feelings, explore characters, and examine the narrative from differing perspectives (Sawyer & DeZutter, 2007). A play would offer an exemplary way of engaging in Richardson's 'crystallization' in order to explore the different faces, experiences and insights of a war (Denzin & Lincoln, 2003).

This was simultaneously a study of the 'other' and of the 'self.' The 'other' lay in our distance over time and my inexperience of war; the 'self' lay in the familiarity of space, and the bonds and common values of family. I knew most of the people who appear in this play, even though I did not know them in the context of this story. It is also a study of the relationship *between* the 'other' and the 'self.' It was this reciprocity, and the insights that it might provide, that I was most interested to explore: I struggled to determine how I knew what I knew, and how I could represent that 'knowing' in ways that others could also 'know.' The challenges in writing this play were always around authenticity, honesty and compassion. How could I use a person's story and respectfully turn it into data? How could I translate my father's story as a small orphaned boy into a story that resonates with our national history, and which opens that history up to scrutiny? How could I respond compassionately to the things that mattered to my father, when different things were important to me? As my father's daughter, how much of this story was mine?

At times there was a conflict, and at other times, synergy, between my processes as an historian, as a social researcher, a playwright and a daughter. Simultaneously exploring the 'other' and the 'self,' the crystalline process for research became an expression of self, as well as of the research: self as researcher, writer, documenter, and daughter whose perceptions reflected and refracted history, story, community, family and identity.

As a way of beginning to understand these letters and build meaning from them, I used strategies from Augusto Boal's Theatre of the Oppressed (Boal, 1979). Theatre of the Oppressed creates a theatrical space in which oppression can be explored, understood and overturned, where people who are oppressed can be heard. War creates oppression in many spaces. I was concerned with my father's oppression as a child - paedophilia, powerlessness, disenfranchisement in making life decisions, emotional isolation and the use of child workers all arose in the conversations with my father when researching this play. But the adults in my father's world were also oppressed by circumstances, expectations and fears. Consequently, I employed Boal's Rainbow of Desire, one form of Theatre of the Oppressed, as an interesting and effective way into creating and communicating meaning (Boal, 1995). In The Rainbow of Desire monologues are expressions of longing. This offered me a way into reading the letters, when I began to listen for the 'heart's desires' they expressed, or to which they responded. It gave me a way to create scenes which could be 'read' emotionally by the audience.

The processes of research, reflection and refraction, which underpin every aspect of the creation of the play, can be explored through the analysis of a scene.

The following extract is based on one of Bill's most stressed letters. It is the first he wrote after his pivotal experiences on Greece and Crete, and it is addressed solely to my father, who had just turned 11. In this scene my interest is in trying to unpack the levels of stress, fear and shock that underlie the text, without adding to the letter itself. I have done this by creating a duologue between Ron's reading of the letter and Bill's feelings while writing the letter. In many ways, Bill's words are a monologue, a rainbow of desire, juxtaposed against his presentation of himself through the letter. Ron's reading of the letter, takes the words of the father and has them spoken by the son, freeing Bill's words to be a genuine expression of his fear, anxiety and relief. When the character of Ron reads Bill's words, the nuances of Bill's letter are altered, by Ron himself, and by the audience's response based on their knowledge of the play and on their own experiences of fathers and families. There is a question here about how we read and how we understand. In this scene I sought to exploit this ambiguity in order to create a pathos with which I hope the audience can connect. Of course there is an added, implicit layer: this is my understanding of Bill's experience, shaped as it has been by the intervening 70 years, by historical knowledge and by later events, and simultaneously reflects my perceptions of my father as a boy and my knowledge of him as a man. Perhaps the most important subtext here concerns the idea of 'fatherhood.' In the play Bill is the father, but in my life experience it is Ron who is the father. As writer and researcher it is impossible to remove my knowledge as daughter.

It is not my intention to 'put words in Bill's mouth'; rather, I sought to present this script as an analysis of the data, expressed in human terms in order to create an emotional as well as an intellectual connection with the imagined audience. It was, however, my intention to 'take the words out of Ron's mouth.' By reading Bill's letter aloud, Ron is effectively silenced – his own words, his own thoughts and fears, are not heard. This was a specific choice in order to reflect the ways in which children are unheard, silenced, particularly in times of trauma. Often, when preparing this play, my father would say that he didn't feel anything, that he was 'cold' or that he didn't remember how he felt about major events. Yet the stories he tells of his responses (e.g. lying in bed and smelling his father's hair oil, and planning ways to stow away on board his father's ship) speak of repressed feelings, the creation of an 'exclusion zone' around the expression, and therefore the feeling, of emotion. Ron's desires are unheard. Ron reading his father's letter aloud is a strategy to communicate this silencing to the audience, and to create a sense of the way in which Ron was required by circumstances to absorb trauma with few opportunities to express it, and with no frameworks to analyse or understand it.

The audience are not merely viewers, but participants in creating the emotional and psychological world presented in the play: everyone has a father; everyone has different ways of knowing pain, loss and trauma. This scene is an attempt to link the audience's experiences of life with those of the people in the story (Saldana). Although a number of dramatic strategies have been used, I hope the effect is to generate a way of 'knowing' and expressing knowledge, and of prompting others to engage in that 'knowing,' or to find that 'knowing' in themselves.

[The Morgan's front room. Ma and Eileen are listening to music on the radio and both are knitting; Ron enters with a letter from Bill]

[Ron reads the letter aloud. Bill is writing the letter and comments on his writing]

Ron: I've got a letter – it's a very exciting one. Dad's been in Greece! And Crete! But he's all right! Dad's all right!

[Ma turns off the radio and she and Eileen prepare to listen, saying things like Crete? Oh my goodness, Crete was a disaster … Ron settles down to read aloud. Ma and Eileen's involvement in the letter is reflected in the speed and intensity with which they knit. Ron's reading has improved but he will occasionally stumble – he reads with childlike enthusiasm and expression – quite innocent of the horror he is conveying. Eileen and Ma respond to the sense of the letter and seek further information from the atlas, etc.]

Ron: [reading] My Dear Ron, and all the folks at home,

This is for me a very happy moment, for once again I am in the position to be able to write to you once more, it is two months since I wrote to you last, and that was from Egypt, since then a lot has happened, but it is not my intention to write to anyone and say what we have experienced, other than that I am poorer today than ever before.

Bill: How can I tell my son, my little boy? How can I tell him what has happened? And if I don't, if all I talk about, merrily, is what he got for his birthday, then what does that mean when I come home? When we have to start living again, and he has no idea, no idea of the man I will be and the things I have seen and the things I have done? Where are we then?

Ron: [reading]I happen to be amongst those boys that went to Greece, I was there for two weeks, two weeks that I will never forget, then on our way back our ship was bombed and wrecked, the navy again came to our assistance and landed us on the Island of Crete, then the Germans came and now I am back again somewhere in Palestine, our last week on Crete was a real nightmare, and now we, who were lucky enough to get away, talk and retell over our experiences, we don't know yet how many of our mates were left behind and taken Prisoner of War.[during this Eileen gets out the atlas and looks up the locations]

Bill: But he is a child – just a little boy – he doesn't need to know – shouldn't know. Can he imagine? What do I say – to set his mind at rest and yet to tell him not to be at rest, that today I'm fine but tomorrow I might not be? How do I prepare a child for death – who's already seen too much death, been too close to death, been thwarted by death – how do I tell him that here I am, putting myself in death's way, without scaring him? Perhaps it is right to be scared. Right, but not fair. Not for a child.

Ron: When the ship went down I lost everything I possessed, at the time of writing I have a shirt, pants and a pair of worn out boots but we are expecting an issue of clothing any day. The navy has been wonderful, three times they came to our assistance and saved us from peril; the boys have been like brothers to us, they cannot do enough for us, the Red Cross and the Comfort fund have been very good also handing out … cigarettes, toothbrush and tooth paste, toilet equipment and chocolate which we all very much appreciate.

Bill: He will like me mentioning chocolate. He will think that everything is fine because of the chocolate. He will know I am being cared for. I never knew how much I loved him til this moment – this moment …. How many of my mates had sons, little boys? Oh God! I can't … [but he must, so he picks up the pen again to write]

Ron: I suppose you received my cable to say that I was safe and well, just after I sent that cable the fun began and I was not so safe, but now I can say, for the time being, I am safe and keeping in the best of health.

Bill: Safe. Safe. Oh Ron, be safe.

Ron: How are you all at home? I have not heard from any of you for over two months there must be some letters about somewhere.

Bill: Let's talk about normal things. Life things. Home things.

Ron: How did you spend your birthday? Did Grandma give you the pound that I mentioned in my last letter and what did you procure with it?

[Pauses and thinks, then continues]

How is Lull and Bryan Margaret and Albert, Eil and Jack, Myrtle and family, Lou, Grandma and Grandpa Taylor, Lin, Joyce, Alma and Gordon, Eunice and Nancy?

Bill: Say the names, like a mantra. The people I love, the names I love that roll off my tongue. Like a charm, like a spell. The names will keep me safe. Let's talk of family – my family – and Ron can tell me all about them. That's his job. He can tell me about them.

Ron: I am going to write to them all once I settle down and am my old self again,

Bill: I will – I will do that Ron, but first I am writing to you. Because you are my son. And even though you are a child, you are closer to me than anyone. I love you most, you are the most precious in my life. I will write to them, but let me first touch you through this letter. I write the letter, I touch the paper, you open the letter, you touch the paper, you read my thoughts. Let me first write to you. And, then when I am my old self, not this new self who writes now, in panic and fear, but my old self, the one they know, then I will write. But you need to know this new me – because I love you most, because you need to know. I have changed.

Ron: I have lost a lot of mates but things will right themselves within a day or two.

Bill: I have changed. It will be all right. But things have changed. I have changed. Things will right themselves. I need things to be right with you Ron. I need you to know.

Ron: I would like you to send me some snaps of you and home, apart from a couple I kept in my pocket I lost the rest; tell Lou and grandma that I still have the pen, but like a good many of the boys we had no ink, but things are becoming much better, we are to receive some pay Friday so that will be just like old times.

Bill: Send me some pictures – do something. This is a war of inaction – we spend months doing nothing and then we are swept up in battles that are not our own and we find ourselves overwhelmed with action. I know why they call it action. Action and inaction. And what can my son do? – let him act. Let him find photos – pictures that bind us. And let him talk to the others – let them reassure him. I still have the pen – we can still write, I am only so far away as a letter. Different if I were dead. A pen cannot cross the gulf between life and death ....

Ron: Write to me often, won't you, tell me about Rees, what unit he is in, and where he is, I am very anxious about him, don't forget to send me the football news, starting from the beginning of the season as I have not heard much news; would you cut out the weights for the National and send them also any First Aid books (St John's) as all mine are gone.

Bill: Here are some jobs for you to do – some reasons each day to think of me. I worry about my little brother – where is he now. After the bloody mess of Greece and the disaster of Crete – so many dead, or gone –or dead – what will they do with Rees? Write to me. It doesn't matter whether or not we get the letters. Write to me. It is writing that keeps us alive for each other. When you write to me I am real for you. Write to me often.

Ron: Tell Grandpa and Grandma Morgan not to worry about me as I am in the pink and on no account must they send anything to me, I will be writing to them very shortly.

Bill: I am in the pink. I'm breathing. I'm not in German hands. I'm in the pink. Yes, I am, I have nothing, but don't send me anything. I will lose it all – it will never arrive. They have no idea where I am or where I'm supposed to be. It's a bloody mess – how will anything they send find me here – wherever I am. Send nothing.

Ron: Well, dear Ron, I must conclude now.

Bill: There's nothing more I can say without saying everything. Don't worry – worry will not save me – will just make you miserable. Do not worry. I'm in the pink.

Ron: and trust this letter [finds] you all in good health.

From your loving Dad, Bill.

Bill: Your loving Dad. Your very, very loving Dad.

I have no doubt that Bill loved his son, my father. His letters are affectionate, offering advice, guidance, and reflection on their lives and the pain of separation from each other, and from Joyce, Bill's deceased wife and Ron's mother. However, I hope the play also offers an 'empathetic and embodied engagement of other ways of knowing' (Alexander, 2008, p. 77). Amongst these 'other ways of knowing' is the heart as opposed to the rhetoric. In the course of researching Bill's activities during the war I came across a pastel portrait done of him by a war artist during his time in training in 1944 at Herberton. The accompanying note states:

> Portrait of Sergeant William John "Mick" Morgan (VX2495), a nursing sergeant in 2/2 Australian Field Ambulance. Sgt Morgan's conscientiousness and interest in the treatment of patients under his care earned him the respect of many in the Brigade. He provided service in the Middle East, the Western desert, Greece, and Crete; and he saw action in Syria and Ceylon as well as Wau, New Guinea as a Nursing Sergeant (AWM/ART21414).

This description of my grandfather as 'conscientious' is one that bounces like light off a crystal. It is consistent with the received rhetoric about members of the second AIF who left their families in an act of ultimate sacrifice in order to defend their homes, their loved ones and their values. But when I write as a daughter, rather than as an historian, I struggle: how does a conscientious man leave a 10 year old as an orphan in order to go to war? What does this word 'conscientious' mean? In his letter written the night before his first entry into battle, the attack on Bardia, Bill writes to his mother:

> For all that can be said against it, I still maintain that this war is a very good thing; every individual is having the chance to give and dare all for his principle like the martyrs of old; however long the time may be, one thing can never be altered – I shall have done my duty like a true Australian, nothing else matters one jot, nor can anything ever change it (Morgan, late 1940).

The letter continues in a similar vein – an eloquent expression of Bill's perception of his place in the world – concluding 'we will live again in peace and freedom, and having paid our contribution to the motherland, our lives will not have been in vain.' These fine words are absolutely consistent with the values of the time, and the presentation of the motivation of people who leave families to go to war, even today. The Historian within me acknowledges that the Second World War was fought between factions representing ideals, and that for war to continue those ideals had to be manifest in the actions of individuals, that notions of sacrifice, freedom and peace had to be objectified in a way that made them worth dying for. Within this context, Bill was indeed a conscientious man – in his actions, his acceptance of his role, and the commitment with which he undertook it. With the power of hindsight, however, Bill's values and the values imposed on him, sit uncomfortably on my page. Bill believed he would be 'called up' as a single man because the concept of a single father was unimaginable. He was denied the idea of choice by the dominant social constructs. When Bill returned from war, my father was 16, a young adult. Bill married again, a much younger woman, and although the letters continued, about football, racing and socks, my father never lived with him. Bill had a second family the same age as me and my siblings. After the war my father never knew Bill as a father. The relationship between father and son was the greatest casualty of Bill's war. As my father reflects, "War was conducted on a variety of levels – and this one Bill lost." To accept this idea of Bill as 'conscientious,' therefore, recognises the quality and care of his work in the field ambulance, but privileges this story of war over any other. It privileges the social mores that effectively denied choice, and the oppression of Bill as a father, at the cost of the story of loss embedded in Bill's and Ron's lives. Why we choose, as a society, to validate and then canonise some stories over others has to do with power, social constructs and frameworks of thinking (Kincheloe and Steinberg). One of the functions of this play is to claim back other ways of knowing that time and to give audiences a way to 'come to know [their own] culture differently' (Alexander, 2008, p. 77).

Somewhere between the history and the story, or perhaps within them both, lies the truth of Bill's experience and of Ron's. Meaning is constructed somewhere between the origin and the reception, somewhere in the letter, between the writing and the finding. I hope that in writing this play I have been as responsive as possible to both these stories, and have brought them together in a way that allows those outside the story to connect with it, and to explore the emotional and historical landscapes in which it exists.

REFERENCES

Abbott, Harold. Portrait of William 'Mick' Morgan, found at: http://cas.awm.gov.au/item/ART21414.

Alexander, B.K. (2008). Performance ethnography. The reenacting and inciting of culture. In N.K. Denzin, & Y.S. Lincoln (Eds.), *Strategies of qualitative inquiry* (pp. 411-432). Thousand Oaks, CA: Sage.

Beevor, A. (2005). *Crete. The battle and the resistance*. London: John Murray Publishers.

Boal, A. (1979). *Theatre of the oppressed*. London: Pluto Press.

Boal, A. (1995). *The rainbow of desire*. London: Routledge.

Commonwealth of Australia (2001). *A great risk in a good cause. Australians in Greece and Crete April-May 1941*.

Costa Rica, photograph of the sinking, found at: http://cas.awm.gov.au/item/P05094.001.

Denzin, N.K., & Lincoln, Y.S. (2008). The discipline and practice of qualitative research. In N.K. Denzin, & Y.S. Lincoln (Eds.), *Strategies of qualitative inquiry* (pp. 1-32). Thousand Oaks, CA: Sage.

Frean, Rev W. (1949). *Star of the sea over Crete*. Pennant Hills: Majellan Press.

Jackson, A. (2008). *Theatre, education and the making of meanings*. Manchester: Manchester University Press.

Kincheloe, J.L., & Steinberg, S.R. (1998). Students as researchers: Critical visions, emancipatory insights. In S.R. Steinberg, & J.L. Kincheloe (Eds.), *Students as researchers. Creating classrooms that matter.* London: Farmer Press.

Morgan, Bill (1940). Letter, undated late 1940, Private collection.

Morgan, Bill (1941). Letter, dated 8/10/1941, Private collection.

Morgan, Ron (2011). Interviews conducted during 2011. Author's notes.

Robinson, C. (1991). *Journey to captivity*, Canberra: Australian War Memorial.

Saldana, J. (2008). Ethnodrama and ethnotheatre. In J.G. Coles, & A.L. Knowles (Eds.), *Handbook of the Arts in Qualitative Research.* Thousand Oaks, CA: Sage.

Sawyer, R.K., & DeZutter, S. (2007). Improvisation: A lens for play and literacy research. In K.A. Roskos, & J.F. Christie (Eds.), *Play and literacy in early childhood. Research from multiple perspectives*, 2nd ed. London/ New York: Lawrence Erlbaum Associates.

Swain, Bruce T. (2001). *A chronology of Australian armed forces at war 1939-45*. Crow's Nest: Allen and Unwin.

Mary-Rose McLaren
School of Education
Victoria University
Melbourne, Australia

SARAN STEWART

PROBLEMATIZING RACISM IN EDUCATION

A comparative analysis of critical race theory and postcolonial theories using autoethnography

INTRODUCTION

Critical race theory (CRT) and postcolonial theories (PCT) are kindred in principles, and challenges ahistoricism by providing methodological constructs to analyze racism[i] in education in both a historical and contemporary context using interdisciplinary methods (Andreotti, 2011; Ghandi, 1998; Johnston, 2003; Ladson-Billings & Donnor, 2005; Solorzano, 1998; Spivak, 1988). Towards this end, both epistemologies can be used as methodological frameworks for examining oppression through a racialized lens (Ladson-Billings & Donnor, 2005). Critical race theory sets the stage in which autoethnographic methods can expose the ways in which racism operates in daily life (Denzin, 2005). Comparatively, scholars have argued that the ongoing debate surrounding postcolonial theories as a measure of research is knowing how to differentiate the subject from the object of analysis (Kincheloe & McLaren, 2005; Smith, 2005). Furthermore, postcolonial theories makes *visible* and *names* the history and legacy of European colonialism (Rizvi, Lingard, & Lavia, 2006), thereby problematizing the effects of postcolonial rule.

Critical race theory positions racism as endemic to American life (Matsuda, Lawrence, Delgado & Crenshaw, 1993). In comparison, postcolonial theories situate racism at the epicenter of colonialism and the lasting effects of colonial power in postcolonial countries (Fanon, 1961; Johnston, 2003; Spivak, 1988). Through this scholarly paper, I will comparatively analyze critical race theory and postcolonial theories using personal narratives, in hopes of making the personal political (Holman Jones, 2005). As a measure of authenticity and transparency in discourse and dialogue, I will use autoethnography to present myself as both the researcher and the research subject (Holman Jones, 2005). Towards this end, I will provide a review of critical race theory and postcolonial theories in education, and comparatively analyze my educational experiences in Jamaica and the United States. Consequently, I will answer how critical race theory and postcolonial theories can deconstruct racialized experiences in education as a constructive measure to galvanize change and social activism.

M. Vicars et al. (eds.), Discourse, Power, and Resistance Down Under, 199–213.

CRITICAL RACE THEORY

As a "legal framework rooted in the civil rights movement" (ASHE, n.d.), critical race theory (CRT) evolved from its roots in jurisprudence to a race-based epistemology to analyze and deconstruct racial inequities in the United States (Harper, Patton & Wooden, 2009). As an interdisciplinary approach, CRT can be applied to many disciplines of law, education, sociology, history, anthropology and "women's studies to advance and give voice to the ongoing quest for racial justice" (Harper et al., 2009, p. 390). Specifically to education, scholars have argued that CRT problematizes aspects of curriculum, pedagogy, institutions, schools and classrooms that perpetuate the rationale, causes and consequences of educational inequalities (Ladson-Billings, 2009; Sleeter, Delgado & Bernal, 2003; Yosso, 2002). Towards this end, Yosso (2002) argued that "racism and its intersections with discrimination based on gender, class, language, and immigration status informs curriculum in both macro and micro ways" (p. 93) leading to the misappropriation of a "race neutral" and "colorblind" curriculum. Concurrently, critical race pedagogy challenges the traditional dominant, normative ideologies of teaching, and makes space for an interdisciplinary approach to experiential teaching and learning (Solórzano & Yosso, 2001). Furthermore, critical race pedagogy is centered on the lived experiences of students of color and how those experiences provide teachable moments in the pursuit of critical consciousness (Sleeter & Bernal, 2003).

Critical race theory plays a poignant role in this analysis through the six tenets that best supports the comparative analysis. Adapted from Matsuda, Lawrence, Delgado and Crenshaw (1993), the six tenets are: (1) racism is endemic to American life and it's educational system; (2) race neutrality, objectivity, colorblindness and meritocracy are permeated throughout the institutional structure of education; (3) racism is a contributory factor to group advantage and disadvantages; (4) CRT allows for recognition of experiential knowledge of marginalized people and communities such as students of color; (5) CRT is an interdisciplinary approach that is adaptive to many contexts including but not limited to education; and (6) CRT provides a commitment to social justice for the broader goal of ending oppression. I selected CRT because it is an evolving methodological approach that "challenges liberalism and the inherent belief in the law to create an equitable just society" (Morfin, Perez, Parker, Lynn & Arrona, 2006, p. 252). Furthermore, a CRT analysis provides an interdisciplinary approach to link postcolonial theories as progressive measures of activism.

POSTCOLONIAL THEORIES

Postcolonial theories (PCT) are an amalgamation of several critical ideologies that form a complex analytical approach to problematize the effects of colonial subjugation and the marginalization of indigenous peoples (Ghandi, 1998; Said, 1978; Spivak, 1988; Smith, 1999; Tikly 2001). Furthermore, postcolonial theories foregrounds racial differences in the political, social, economic, and cultural relationships between the Global North or developed countries and the

Global South or developing countries (Fanon, 1961; Johnston, 2003; Spivak, 1988). Towards this end, scholars have used postcolonial theories to problematize the representation of developing countries and issues of power, voice and cultural subordination as it relates to a Eurocentric, normative, dominant ideology (Andreotti, 2006; Rizvi et al., 2006).

The uniqueness of this approach lies within its paradoxically simple but complex definition, which "refers to that, which has been preceded by colonization" (Johnston, 2003, p. 35). However, scholars such as Dirlik (1994) position the term postcolonial to have a simultaneous dual meaning to global conditions after colonization, and an "epistemological and psychic orientation that are products of those conditions" (p. 332).

Accordingly, the emerging discourse on the reconceptualization of colonialism through re-narrativization is attributed to postcolonial theories (Tikly, 2001). Essentially, this allows for a deconstruction of "the collective memory of imperialism" (Smith, 1999, p. 1) and the "reclamation of knowledge, language and culture" (Smith, 2005, p. 88). As such, I will use postcolonial theories in education to reveal the legacies of colonial education and the similarities drawn from critical race theory.

Postcolonial theories in education

As an interdisciplinary approach, postcolonial theories can be used to link other academic disciplines such as education to galvanize progressive activism to challenge the impacts of colonial rule (Johnston, 2003). Postcolonial theories in education are an amalgamation of several counterhegemonic ideologies that challenges traditional construction of knowledge (Andreotti, 2011).

This contextualization of PCT in education is a meta-theory of epistemological approaches and frameworks used to legitimize and reconstruct knowledge in schools, colleges and universities. I use this meta-theory dichotomously to 1) deconstruct my experiences in Jamaica's education system and 2) establish new ways of knowledge construction and critical pedagogical approaches to galvanize systemic change.

Reddock (1994) critiqued the effects of colonial education as a means to not "liberate the colonized, but rather as the means whereby the values and interests of the colonizers and masters would be internalized by the colonized and perceived as their own" (p. 48). Accordingly, PCT provides a methodological construct to interrupt the "negation of the unrecognized accounts of the postcolonial subject" (Lavia, 2006, p. 189). Thereby, narratives from the postcolonial subject provide an ontological reconditioning of the postcolonial mind that conversely values and chronicles untold histories.

METHODOLOGY

I selected autoethnography as a medium to critically analyze my educational experiences through a racialized lens. Ellis (2004) stated that autoethnography is comprised of "research, writing, and methods that connect the autobiographical and personal to the cultural and social" (p. xix). Autoethnography is a practice of reflexivity at its best where the self is the unit of measure, and there is a concerted effort to continuously *see the self, seeing the self through and as the other* (Ellis, 2004; Alexander, 2005). This dichotomous process of internally and externally critiquing the self is to juxtapose and project the political, cultural and democratic constructs of self in society. Alexander (2005) and Holman Jones (2005) argued that the content and form, however flexible and subjective in autoethnography, reflect rigor, substantive contribution, realistic context, and aesthetic merit.

Accordingly, I selected key representations of my educational experiences in Jamaica and the United States to illustrate moments and legacies of racism. As a measure of comparative analysis, I interwove the principles of both theoretical constructs into my lived experiences to give critical context to microaggressions,[ii] and subjugations of oppression.

I selected and analyzed moments of my educational experiences that best represented racialized occurrences. Coupled with my personal texts, I developed a pictorial vignette of my lived experiences in Jamaica and the United States. At the 2011 Association for Qualitative Research/ Discourse, Power, and Resistance Conference, I presented the pictorial vignette as a part of my findings. Below, I have included screen shots of the presentation to best grant aesthetic merit to the re/presentation of my autoethnography.

OPERATIONALIZING CRT AND PCT IN EDUCATION USING AUTOETHNOGRAPHY

The selection of narratives is a deliberate methodological construct supported by both CRT and PCT to interrupt the "negation of the unrecognized accounts of the postcolonial subject" (Lavia, 2006, p. 189). Additionally, narratives as acknowledged in CRT gives voice to the undocumented history of the silenced and the oppressed, providing a valued context for those who have been silenced. Towards this end, critical race theory and postcolonial theories in education provide a medium to illustrate the endemic levels of institutional racism and the chronicled subjugation of marginalized students.

Deconstructing colonial education through a PCT lens

During the pre-independence era in Jamaica, students such as my father were expected to study the works of William Shakespeare, Charles Dickens, and W.B. Yeats. Most textbooks used in Jamaican schools were written and published in Great Britain, Canada and the United States, having little relevance to Jamaican children (Whiteman, 1994). Additionally, the textbooks serving as the key instrument in curricular instruction mirrored industrialized, Western societies (Whiteman, 1994). The disconnect that the textbooks represented inhibited the progress of student learning. Lavia (2006) described the premise and reason for colonial curricula as:

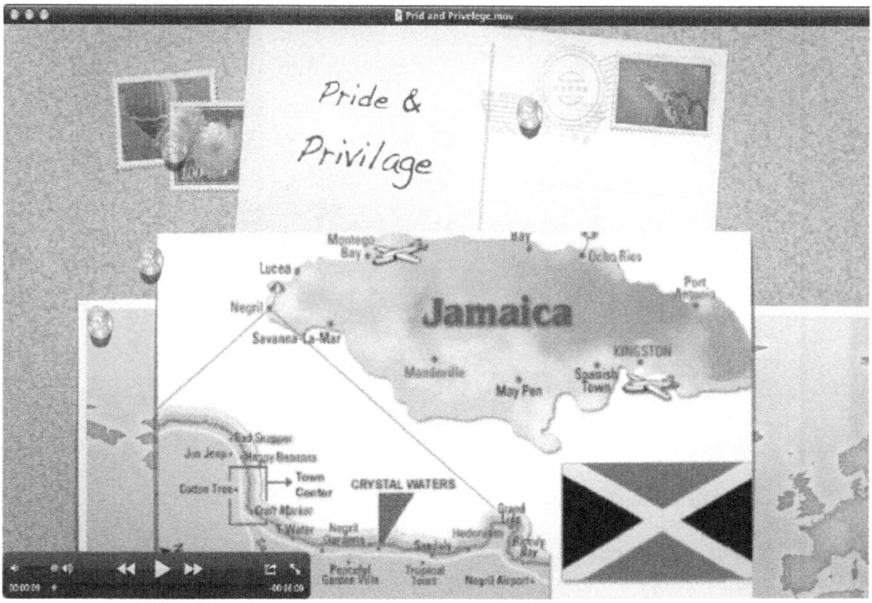

Figure 1. I start the vignette with Bob Marley's "One Love"

> A manifestation of imperialist ideology sought to ensure that the education of 'young colonials' involved mastery of colonial values and ways of being, limiting excising and ignoring any semblance of anything that might perhaps be considered as merging from historical and social truth of the colonized (p. 284).

The physical location of Jamaica and the genetic make-up of its people, prevented the continued acceptance of a normative, European, colonial curriculum. Thirty-years later, my experience was a little different from my father's as it marked the era of change where a pseudo-blended model was adopted but far from perfect.

While in Jamaica, I cannot remember race ever being noted on a birth certificate or government form. However, social class was a measure of where and how you lived. The traditional ghettos of the inner cities were former slave settlements, legacies that still remain present. My primary school was located on the outskirts of the downtown area, a short distance from the penitentiary. However, it was a highly ranked preparatory school, known for academics and sports. The daily lessons at my school resembled a hybrid model of Caribbean and British curricula.

As a grade-school student, I sang the Jamaican national anthem and said the national pledge every morning. However, I was taught about parliament both in Jamaica and England, so that I could compare systems of governance. I was taught about Jamaica's seven heroes and the fourteen parishes of our country. I was

Figure 2. I used reggae music as the backdrop to frame my resistance and opposition to the colonial legacies of education

expected to speak the Queen's English and not my grandparent's patois. I studied the difference in value between the Jamaican dollar and the British pound, always establishing that the Jamaican dollar had a weaker exchange rate. Reflecting now, I see how the subliminal image of a weaker monetary value reinforced my identity as a Third World citizen.

In high school, my English curriculum was a mix of American literature, Greek mythology and Shakespearean plays. Although, my English literature syllabus included Caribbean novels, they were few and far in between. To illustrate the norm of the Western-dominant curriculum, the island-wide, 1998 Caribbean Examination Council (CXC) English Literature Examinations focused primarily on A Raisin in the Sun and Shakespearean plays. I remember having a difficult time reading Shakespeare because of the complete disengagement of language, location and time period. I yearned for books by Louise Bennett, Jamaica Kincaid and Kamau Brathwaite that demonstrated the use of patois and the vibrancy of the Caribbean culture. The omission of Caribbean texts in the curriculum to students, can be tied to the "history of racial exclusions … by the embedding of racialized distinctions into ordinary processes of reasoning and into various [deceptions] of morality" (Johnston, 2003, p. 29). Not having these texts as my required reading, forced me to read them outside the classroom letting me know at the time, that they were not valued for in-class curriculum.

As a repository of the "banking system of education" (Freire, 1993), I struggled in classrooms of forty-students, foreign curriculums and strict discipline, where I was expected to be silent and take notes. After years of silencing, I was enrolled in Mr. Bell's extra lessons in third form, where I was engaged in dialogical discourse and critical thinking. Extra-lessons are essentially private tutoring and/ or additional academic lessons outside of the regular school curriculum (Lochan & Barrow, 2008). There, I began to regain my voice. As a result, I was challenged to go against the status quo and engage in different learning styles and pedagogical reflexive approaches, where I was encouraged to embrace the role of teacher and student concurrently. As a resulting factor, I now identify as a postcolonial-emancipatory-rebel that rejects the banking-system of education, and yearns to radically reform Jamaica's education system.

Deconstructing American education through a CRT lens

I can still feel the dread of despair in the pit of my stomach as I remember my father unloading the last bags out of the rented minivan parked in front of my dorm. I had arrived at my new school shortly after my seventeenth birthday. It was a preparatory, boarding high school in the New England area that rivaled other elite, presidential schools. Even though I had graduated form high school in Jamaica, I was 15 at the time, and my parents learned that it would be easier for me to receive a scholarship to an Ivy League university if I attended a prestigious New England preparatory school. The dorm was a two-bedroom unit attached to a small, cottage-style house on top of a hill, a half-mile from the campus. I remembered looking around at the abundance of pine trees and thinking that I was not in Jamaica anymore.

Figure 3. I selected this image to clearly depict the stark contrast between the tropical paradise I left and the cold reality I had entered. Additionally, this was my first encounter with snow

Figure 4. As the only Black female in my junior and senior years, I experienced recurrent feelings of being ignored, islated, lost and silenced

As Tyack (1995) stated, my differences were ignored daily; I was grouped with the other nine students of color, of which three boys were also from Jamaica (recruited on a football scholarship). The fact that I was Jamaican meant that I was just a token; I was foreign enough to be exotic but too "third world" to be accepted. I was recruited for diversity purposes and was the only Black female in both my junior and senior years. I felt like a Black rag doll in the Barbie collection aisle of an American toy store, completely out of place.

I would later learn that my high school diploma and transcripts from Jamaica as well as my Caribbean Examination Council (CXC) results were never converted to the American equivalent. As such, I was placed in remedial classes without the option of testing out. My English teacher neglected to consider that my words and constructed sentences were based on the British curriculum and therefore, I was neither dyslectic nor incapable of spelling words such as *colour*. I was repeatedly denied entry into calculus even though I had passed with distinctions, the equivalent of advance placement (AP) calculus in Jamaica.

A newly hired, young teacher, Ms. A., not-yet socialized and accustomed to the norms of the preparatory school, spoke with me at lunch one day and heard my plea to test out of finite math. She tested me and immediately referred me to be accepted to her pre-calculus class. However, the math department chair, who was about to celebrate his twentieth anniversary at the school denied her referral. He, being my finite math teacher, told her that I was unable to master the use of the TI-

93 calculator and as such would not be able to succeed in pre-calculus. He was right about the calculator because in Jamaica, I never had to use a TI calculator. Only basic calculators were allowed in CXC exams. As a student in Jamaica, I was taught how to graph on graphing sheets, and learned how to integrate and differentiate on paper not on calculators.

The repercussions of this cyclical abuse of hierarchical, academic power hindered my academic progress. I felt raped of my identity; I felt my sense of self being ravaged day after day. Whether it was constantly defending my academic potential or answering nonsensical questions such as *are there roads in Jamaica; do you live in trees in Jamaica; can you repeat that, I cannot understand you*; I felt marginalized and silenced. I would do the bare minimum to stay under radar; after all, I was never expected to do any better. Upon returning for my senior year, I wrote a compelling proposal to the headmaster to allow me to dance at the nearby state college. He agreed and through dance, I survived my senior year.

Figure 5. I was asked to dance for the annual Fall Parent's Weekend and as much as I was being used as a token, is as much as I used to dance to survive

My struggle with U.S. educational racism continued at the postgraduate level. At the completion of my first year as a postgraduate student, I saw how faculty embraced *White Comfort*, silenced the race talk, used sanitized language to unacknowledged differences, and then perpetuated like Wall Street, the false economies of niceness (Galman, Pica-Smith, & Rosenberger, 2010). Towards the end of my first year as a postgraduate student, I wrote a reflective brief in the form of a poem, illustrating the emotional and physical toll that I had experienced:

> I am exhausted, mentally, physically and emotionally
> Passing the point of delirium
> Resolved with the constant hallucinations of talking papers and dancing APA manuals.
> I feel raped metaphysically
> Forced to look inwardly, and question my most salient identities
> Who am I? Black? Jamaican? Scholar? Mother?
> all of the above?
> There is an invisible mirror before me
> On the train, in the classroom
> The reflection from my IPad
> on my Apple screen.
> Reflecting parts of myself that are
> Constantly critiqued or rejected
> Pending on the space I occupy in that moment
> of authentic transparency
> For as much, as I am transformed
> Is as much as I have remained the same
> Because, still outwardly
> I will forever remain, just another Black face.
> In the classroom … on the train … pushing a stroller … tired, with glossy eyes

ANALYSIS

Education has the dual role of perpetuating racism as well as reducing racism's longitudinal effects. However, when institutional racism is propagated, embedded and perpetuated in institutions of education, the effects are inextricably damaging on society as a whole because education is arguably considered a social process vehicle that can be used to propel change or reinforce ignorance (Gillborn, 2010; Lopez, 2003). Through a postcolonial theoretical lens, I reflect on my role as a transnational; I am both native and foreign at the same time, playing the double consciousness role, unable to separate myself from the *other*. As the other, I experience as Delgado and Stefancic (2001) termed a differential racialization whereby my identity reflects "a notion of intersectionality and anti-essentialism" (p. 9). Thereby, my identity as a scholar, maps an epistemological journey through time and space hailing multiple identities in any given moment.

Commencing in Jamaica, the journey started as a mimicked cloning of British morals and culture through educational legacy. Jamaica having gained independence from Great Britain in 1962 inherited and still practices many of the traditions, laws and customs of England. Between 1984 and 1999 (when I attended school in Jamaica), Jamaica's education system was transfixed by the instrumentalist approach to literacy where literacy was characterized by "mindless, meaningless drills and exercise" (Macedo, 1993, p. 37). Essentially, I rejected the banking system of education, always questioning the relevance of a British curriculum in a majority Black developing country; knowing that I did not speak Shakespearean English and saw coconut trees outside my classroom windows not Maplewood and pines.

It wasn't until 1995, when I attended Mr. Bell's Extra Lessons that I engaged critically in a dialogic discourse of shared knowledge. There is where I first experienced education as the practice of freedom (hooks, 1994). Mr. Bell's liberating and empowering transference and co-construction of knowledge provided avenues for reflexivity and experiential learning. In the very first class, he challenged me to become a change agent and social activist, reminding me that I was the youth of the country and had a responsibility to make changes. I no longer spoke out of turn, I was expected to challenge his style of teaching and reassured that there were many styles, and I could debate each freely without feeling marginalized. It was almost surreal to attend his classes because he provided a safe environment in which we as students were also empowered as teachers. If we found better ways of understanding differentiation and linear equations, he encouraged us to go to the board and explain. He spoke to us in patois and we dialogued with him equally. He was an economics Harvard graduate, able to work anywhere in the country but chose to teach.

Figure 6. It was within this simple home that my movement to critical consciousness began

Comparatively, my experiences in the United States education system were a dance of subtle microaggressions and white interest-convergence. The New England preparatory school was the epitome of Hammond's (1995) depiction of a successful scholastic institution; it had the small faculty-student ratio, the "qualified" teachers, the pseudo-family environment, where students ate with faculty, lived with faculty and called them dorm mother and father. However, Hammond (1995) representing a White, normative ideology, neglected to include the voices of the marginalized students of color. My first year at the high school was met with much dismay and utter disregard for my culture, educational achievement and activation of voice. My culture and nationality were ignored daily; I was grouped with the other nine students of color on campus, and treated like an American rather than a Jamaican. My culture was subsequently tokenized and requested for display on special occasions like Fall Parent's Weekend. I was recruited for diversity purposes and was the only Black female in both my junior and senior years. Through critical race theory, I am able to name the microaggressions and recognize the subtle acts of silencing and interest-convergence (Harper, et al., 2009). These endemic roles perpetuate the strategic positioning of racism in American education, that maintains White interest-convergence as a precursor to Black success.

In the United States "students like [me] have to make a difficult choice between advocating for [myself] and other students of color and the price of total exhaustion from constantly having to step into the advocacy and defense role" (Galman et al., 2010, p. 231). Having attended predominantly white institutions since my baccalaureate years to my current enrollment as a postgraduate student, I am familiar with the marginalization of students of color and the constant battle fatigue endured daily in the pursuit of social justice.

CONCLUSION AND IMPLICATIONS

The work of change comes from the work of resistance whereby to address an area of inequity; there must be opposition to the norm. Critical race theory and postcolonial theories center race as the overarching prism in which to deconstruct and counter acts of oppression. Racism is endemic to everyday life, impacting outcomes of marginalized and oppressed populations globally and in multiple sectors including education. As such, there is a call for change, in which historically institutionalized racism in educational institutions can cease to exist with the initial steps to a radical shift in the adaptation of critical pedagogy.

Through the deconstruction of both narratives via a critical race theoretical and postcolonial theoretical lens, the need for transformational education reform in the U.S. and the Global South is vital to the emancipatory liberation and social justice acclamation of marginalized people. From the dual perspective of an aspiring Caribbean scholar studying in the United States, I will use this paper to purposefully fill the gaps of Caribbean postcolonial literature and add to the existing amalgamation of critical ideologies in the field of race and racism.

To offset the endemic power struggle between the marginalized and the powerful, the colonized and the colonizers, I look to a Freirean[iii] approach of critical pedagogy where the "teacher's thinking is authenticated only by the authenticity of the student's thinking" (Freire, 1993, p. 77). Critical pedagogy provides a tool in which to generate an emancipatory position that would elevate the mind of the subjugated to the mind of the liberated (Lavia, 2006). I have looked comparatively at the principles of critical race theory and postcolonial theories, and seen where the intersectionality of personal narratives can inform the initial stages to adopt a practice of critical pedagogy. Similar to my experiences at Mr. Bell, critical pedagogy requires an "emancipatory interest and … a pedagogy of hope which draws upon evidence provided [from] within"(Lavia, 2006, p. 291). Coupled with the tenets of critical race theory and principles of postcolonial theories, a reflective practice of critical pedagogy can transform traditional, normative, Westernized education to empower and enfranchise the marginalized.

NOTES

[i] I use the term racism throughout this paper to refer to an epistemology and system of advantage based on the social construction of race by dominant, normative, Western ideology (Tatum, 1992).

[ii] I define microaggressions as the negative actions, innuendos and offensive mechanisms against victims of racism (Yosso, Smith, Ceja, & Solorzano, 2009).

[iii] Referring to Paulo Freire's (1970, 1993) term of critical consciousness and the need to reform traditional education to reflect critical pedagogy.

REFERENCES

Alexander, B. (2005). Performance ethnography: The reenacting and inciting of culture. In N. Denzin & Y. Lincoln (Eds), *The Sage handbook of qualitative research* (3rd ed., pp. 411- 442). Thousand Oaks, CA: Sage Publications.

Andreotti, V. (2006). The contributions of postcolonial theory to development education. *Development Education Association*, 1-12.

Andreotti, V. (2011). *Actionable postcolonial theory in education*. New York: Palgrave Macmillan.

Association for the Study of Higher Education. (n.d.). Conceptual approaches. *ASHE Higher Education Report*, 17-27.

Bonilla-Silva, E. (2003). *Racism without racists: Color-blind racism and the persistence of racial inequality in the United States*. Maryland: Rowan & Littlefield Publisher.

Delgado, R., & Stefancic, J. (2001). *Critical race theory: An introduction*. New York: New York University Press.

Denzin, N. (2005). Emancipatory discourses and the ethics and politics of interpretation. In N. Denzin, & Y. Lincoln (Eds), *The Sage handbook of qualitative research* (3rd ed., pp. 933-958). Thousand Oaks, California: Sage Publications.

Dirlik, A. (1994). The postcolonial aura. Third world criticism in the age of global capitalism. *Critical Inquiry*, *20*, 328-356.

Dobson, A. (2005). Globalisation, cosmopolitanism and the environment. *International Review*, *58*(1), 5-25.

Ellis, C. (2004). *The ethnographic I: A methodological novel about teaching and doing autoethnography*. Walnut Creek, CA: AltaMira.

Fanon, F. (1968). *The wretched of the earth*. London: Penguin Press.

Freire, P. (1993) *Pedagogy of the oppressed* (M. Ramos, Trans.). New York: Continuum.

Galman, S., Pica-Smith, C., & Rosenberger, C. (2010). Aggressive and tender navigations: Teacher educators confront Whiteness in their practice. *Journal of Teacher Education, 3*, 225-236. doi: 10.1177/0022487109359776.

Gandhi, L. (1998). *Postcolonial theory: A critical introduction.* New York: Columbia University Press.

Gillborn, D. (2010). The colour of numbers: Surveys, statistics, and deficit-thinking about race and class. *Journal of Education Policy, 25*(2), 253-276.

Habermas, J. (1968). *Knowledge and human interest.* Boston: Beacon.

Hammond, L. (1995). Restructuring schools for student success. In A.H. Halsey, H. Lauder, P. Brown, & A. Stuart Wells (Eds.), *Education: Culture, economy & society* (pp. 320-332). New York: Oxford University Press.

Harper, S., Patton, L., & Wooden, O. (2009). Access and equity for African American students in higher education: A critical race historical analysis of policy efforts. *The Journal of Higher Education, 80*(4), 389-416.

Holman Jones, S. (2005). Autoethnography: Making the personal poilitical. In N. Denzin, & Y. Lincoln (Eds), *The Sage handbook of qualitative research* (3rd ed., pp. 763-792). Thousand Oaks, CA: Sage Publications.

hooks, b. (1994). *Teaching to transgress: Education as the practice of freedom.* New York: Routledge.

Johnston, I. (2003). *Re-mapping literacy worlds: Postcolonial pedagogy in practice.* New York: Peter Lang Publishing.

Kincheloe, J., & McLaren, P. (2005). Rethinking critical theory and qualitative research. In N. Denzin, & Y. Lincoln (Eds), *The Sage handbook of qualitative research* (3rd ed., pp. 303-342). Thousand Oaks, CA: Sage Publications.

Ladson-Billings, G. (2009). Just what is critical race theory and what's it doing in a nice field like education? In E. Taylor, D. Gillborn, & G. Ladson-Billings (Eds.), *Foundations of critical race theory in education* (pp. 17-36). New York: Routledge.

Ladson-Billings, G., & Donnor, J. (2005). The moral activist role of critical race theory scholarship. In N. Denzin, & Y. Lincoln (Eds), *The Sage handbook of qualitative research* (3rd ed., pp. 279-302). Thousand Oaks, CA: Sage Publications.

Lavia, J. (2006). The practice of postcoloniality: A pedagogy of hope. *Pedagogy, Culture & Society, 14*(3), 279-293.

Lochan, S., & Barrow, D. (2008). Extra-lessons: A comparison between "different sides of the track" in Trinidad and Tobago. *Caribbean Curriculum, 15*, 45-69.

Lopez, G. (2003). The (racially neutral) politics of education: A critical race theory perspective. *Educational Administrative Quarterly, 39*(1), 68-94.

Macedo, D.P. (1993). Literacy for stupidification: The pedagogy of big lies. In P. Leistyna, A. Woodrum, & S. Sherblom (Eds.), *Breaking free: The transformative power of critical pedagogy* (pp. 31-57). Cambridge, MA: Harvard Educational.

Matsuda, M., Lawrence, C., Delgado, R., & Crenshaw, K. (1993). *Words that wound: Critical race theory, assaultive speech, and the First Amendment.* Boulder, CO: Westview.

Morfin, O. J., Perez, V.H., Parker, L., Lynn, M., & Arrona, J. (2006). Hiding the politically obvious. A critical race theory preview of diversity as racial neutrality in higher education. *Educational Policy, 20*(1), 249-270. doi: 10.1177/0895904805285785.

Reddock, R. (1994). *Women, labour and politics in Trinidad and Tobago: A history.* Kingston: The University of the West Indies Press.

Rizvi, F., Lingard, B., & Lavia, J. (2006). Postcolonialism and education: negotiating a contested terrain. *Pedagogy, Culture & Society, 14*(3) 249-262.

Said, E. (1978). *Orientalism.* London: Penguin Press.

Sleeter, C.E., & Delgado, Bernal, E. (2003). Critical pedagogy, critical race theory, and antiracist education. In J.A. Banks, & C.A. Banks (Eds.), *Handbook of research on multicultural education* (2nd ed.). San Francisco, CA: Jossey Bass.

Smith, L. (2005). On tricky ground: Researching the native in the age of uncertainty. In N. Denzin, & Y. Lincoln (Eds), *The Sage handbook of qualitative research* (3rd ed., pp. 85-108). Thousand Oaks, CA: Sage Publications.

Solorzano, D., & Yosso, T. (2001). Maintaining social justice hopes within academic realities: A Freirean approach to critical race/Lat/Crit pedagogy. *Denver Law Review*, *78*, 595-621.

Spivak, G. (1988). Can the subaltern speak? In C. Nelson, & L. Grossberg (Eds.), *Marxism and the interpretation of culture* (pp. 271-313). Basingstoke: Macmillan.

Tatum, B.D. (1992). Talking about race, learning about racism: The application of racial identity development theory in the classroom. *Harvard Educational Review*, *62*(1), 1-25.

Tikly, L. (2001). Globalization and education in the postcolonial world: Towards a conceptual framework. *Comparative Education, 37*(2), 151-171.

Whiteman, B. (1994). Education and training partnerships, the 1990s imperatives: Jamaica, The West Indies. *Journal of Education Finance*, *19*(4), 94-98.

Yosso, T.J. (2002). Toward a critical race curriculum. *Equity and Excellence*, *35*(2), 93-107.

Yosso, T., Smith, L., Ceja, M., & Solorzano, D. (2009). Critical race theory, racial microgressions and campus racial climate for Latina/o undergraduates. *Harvard Educational Review*, *79*(4), 659-690.

Saran Stewart
Morgridge College of Education
University of Denver

Lightning Source UK Ltd.
Milton Keynes UK
UKOW06f1109260515

252286UK00001B/35/P

9 789462 090354